KU-796-492

FRINGE PATTERN ANALYSIS

Volume 1163

CONTENTS

(continued)

FRINGE PATTERN ANALYSIS

Volume 1163

UNIVERSITY OF STRATHCLYDE

30125   00736514   0

ML

PROCEEDINGS

SPIE—The International Society for Optical Engineering

Fringe Pattern Analysis

Graeme T. Reid
Chair/Editor

WITHDRAWN FROM LIBRARY STOCK

8-9 August 1989
San Diego, California

Sponsored by
SPIE—The International Society for Optical I

Cooperating Organizations
Applied Optics Laboratory/New Mexico Sta
Center for Applied Optics Studies/Rose-Hul
Center for Applied Optics/University of Ala
Center for Electro-Optics/University of Day
Center for Excellence in Optical Data Proces
Jet Propulsion Laboratory/California Institut
Optical Sciences Center/University of Arizor
Optoelectronic Computing Systems Center/l
    Colorado State University

ANDERSONIAN LIBRARY
★
WITHDRAWN
FROM
LIBRARY
STOCK
★
UNIVERSITY OF STRATHCLYDE

Books are to be returned on or before
the last date below.

LIBREX—

Published by
SPIE—The International Society for Optical Engineering
P.O. Box 10, Bellingham, Washington 98227-0010 USA
Telephone 206/676-3290 (Pacific Time) • Telex 46-7053

Volume 1163

SPIE (The Society of Photo-Optical Instrumentation Engineers) is a nonprofit society dedicated to advancing engineering
and scientific applications of optical, electro-optical, and optoelectronic instrumentation, systems, and technology.

The papers appearing in this book comprise the proceedings of the meeting mentioned on the cover and title page. They reflect the authors' opinions and are published as presented and without change, in the interests of timely dissemination. Their inclusion in this publication does not necessarily constitute endorsement by the editors or by SPIE.

Please use the following format to cite material from this book:
    Author(s), ''Title of Paper,'' *Fringe Pattern Analysis,* Graeme T. Reid, Editor, Proc. SPIE 1163, page numbers (1989).

Library of Congress Catalog Card No. 89-43273
ISBN 0-8194-0199-4

Copyright © 1989, The Society of Photo-Optical Instrumentation Engineers.

Copying of material in this book for sale or for internal or personal use beyond the fair use provisions granted by the U.S. Copyright Law is subject to payment of copying fees. The Transactional Reporting Service base fee for this volume is $2.00 per article and should be paid directly to Copyright Clearance Center, 27 Congress Street, Salem, MA 01970. For those organizations that have been granted a photocopy license by CCC, a separate system of payment has been arranged. The fee code for users of the Transactional Reporting Service is 0-8194-0199-4/89/$2.00.

Individual readers of this book and nonprofit libraries acting for them are permitted to make fair use of the material in it, such as to copy an article for teaching or research, without payment of a fee. Republication or systematic or multiple reproduction of any material in this book (including abstracts) is prohibited except with the permission of SPIE and one of the authors.

Permission is granted to quote excerpts from articles in this book in other scientific or technical works with acknowledgment of the source, including the author's name, the title of the book, SPIE volume number, page number(s), and year. Reproduction of figures and tables is likewise permitted in other articles and books provided that the same acknowledgment of the source is printed with them, permission of one of the original authors is obtained, and notification is given to SPIE.

In the case of authors who are employees of the United States government, its contractors or grantees, SPIE recognizes the right of the United States government to retain a nonexclusive, royalty-free license to use the author's copyrighted article for United States government purposes.

Address inquiries and notices to Director of Publications, SPIE, P.O. Box 10, Bellingham, WA 98227-0010 USA.

D
621·36
FRI

FRINGE PATTERN ANALYSIS

Volume 1163

CONFERENCE COMMITTEE

Chair
Graeme T. Reid, Department of Trade and Industry (UK)

Cochairs
Katherine Creath, WYKO Corporation
Toyohiko Yatagai, University of Tsukuba (Japan)

Session Chairs
Session 1—Phase Shifting Techniques
Graeme T. Reid, Department of Trade and Industry (UK)

Session 2—Image Processing Techniques
Kevin G. Harding, Industrial Technology Institute

Session 3—Applications and Studies of Phase Measurement
Suezou Nakadate, Tokyo Institute of Polytechnics (Japan)

Session 4—New Developments and Applications
Kjell J. Gåsvik, Norwegian Institute of Technology/SINTEF (Norway)

Conference 1163, *Fringe Pattern Analysis*, was part of a four-conference program on Interferometry, Microscopy, and Testing held at SPIE's 33rd Annual International Symposium on Optical & Optoelectronic Applied Science & Engineering. The other conferences were

Conference 1161, *New Methods in Microscopy and Low Light Imaging*
Conference 1162, *Laser Interferometry—Quantitative Analysis of Interferograms: Third in a Series*
Conference 1164, *Surface Characterization and Testing II.*

Program Chair: **Katherine Creath,** WYKO Corporation

INTRODUCTION

This volume contains papers presented at the first SPIE Conference on Fringe Pattern Analysis. This subject is almost as old as interferometry: Thomas Young was one of the first practitioners of fringe analysis when, in the early 1800s, he measured the spacing of "Young's fringes" to determine the wavelength of light.

The wide availability of digital image processing equipment has generated a large and growing interest in fringe analysis during recent years. Technologists from many disciplines have found that by employing automatic fringe analysis, they can extend and enhance the practical use of interferometry. Since fringe analysis algorithms can often be combined with any one of a wide range of interferometers, researchers in different fields of interferometry have collaborated in the development and refinement of their techniques. This conference is a good example of such collaboration. Readers will notice that conference sessions are not defined in terms of interferometric techniques (holographic interferometry, moire interferometry, classical interferometry etc.), but in terms of the image processing techniques that are used to perform the analysis of the interferograms. In any given session, therefore, there are papers that describe the analysis of several different types of interferograms. Some papers go even further and describe the results of research that is not directed at any one branch of interferometry but considers fringe analysis as a subject in its own right.

Graeme T. Reid
Department of Trade and Industry (UK)

SESSION 1

Phase Shifting Techniques

Chair
Graeme T. Reid
Department of Trade and Industry (UK)

Field shift moire, a new technique for absolute range measurement

Albert Boehnlein, Kevin Harding

Industrial Technology Institute
P.O. Box 1485
Ann Arbor, Michigan 48106

ABSTRACT

A new and novel technique is discussed that provides non-contact high resolution absolute measurements with moire interferometry. Field shift moire provides both a course absolute measurement and a high resolution relative measurement similar to a vernier caliper. Combining these measurements yields absolute high-resolution range information.

2. BACKGROUND

A problem that has persisted in moire interferometry over the years has been the lack of ability to obtain an absolute measurement from interferograms with digital heterodyne techniques. A static interferogram suffers from lack of information to distinguish a hill from a valley. By shifting the phase of the fringe pattern, the sign of the slope can be determined, but there remains an ambiguity when the surface in question has a discontinuous jump. To determine the shape of a prismatic surface with block structures as part of the shape, the measurement needs to be absolute, not just relative to connecting points. The reason that one cannot get absolute numbers from the phase shift equation is that the equations rely on the arc-tangent function, which is only continuous over a $-\pi/2$ to $\pi/2$. With two input arc-tangent function, one can determine the quadrant, and therefor the phase over the interval $-\pi$ to π. The inability to determine the absolute phase is termed the modulo two π or two π ambiguity problem. With phase shifting, it is possible to make relative measurements of the points on the interferogram, provided that the surface has no discontinuities greater than the contour interval. A number of methods have been suggested to obtain an absolute measurement[1,2,3] each having their own strengths and draw backs. The technique presented in this paper is hoped to resolve some of the problems encountered by the other methods. In order to assess the practical aspects of the technique, the analysis is performed with the thought in mind that it will be implemented on an 8 bit grey scale vision system.

3. THEORY

A non-telecentric projection moire system is used to project the grating lines onto the object. A similar system is used to image the object onto the reference or sub-master grating. This is a standard projection moire configuration. Typically, one would translate the projection grating to produce a uniform phase shift in the image. For the purpose of this investigation, the entire projection system is translated to perform what is referred to as the "field shift". Because of the diverging beams in the projector and viewing system, the projected grating changes pitch with distance due to the change in magnification. Therefor, the field shift produces a phase shift that is proportional to the height of the object or Z. That is, for any given height level on the object, there is a different grating pitch. Since the distance the grating moves at all levels is made constant by the field shift, the ratio of the amount of shift to the pitch, that is, the degrees of phase of the shift, will be unique to each height on the object. This shift effect is in contrast to typical phase shifting, where only the grating is translated, producing a uniform phase shift throughout the whole field, independent of Z. Therefore, solving for the phase in the field shifted method will provide a unique depth number associated with any Z.

The amount of phase shift imparted in the image during the phase shift procedure is commonly referred to as ß. By solving for ß, or the amount of phase shift at a point in the image, we obtain a coarse three dimensional map of the surface that is proportional to Z. The surface need not be continuous as this is an absolute measure. Now we can also solve for the phase ϕ' of the surface and obtain high resolution data with a 2π ambiguity. Combining these two maps, ϕ' and ß, with the proper equation results in a map of the 2π multiplier or absolute fringe order N. Combining the absolute fringe order map N with the phase ϕ' or fractional fringe map yields an absolute phase map ϕ of the surface. Through transformation equations, the absolute phase map ϕ is converted to an absolute Z map of the surface.

The system configuration used to simulate the problem is illustrated in figure 1. The projector is modeled as a pinhole system located distance "d" from the viewing system, with the principal ray parallel to the Z axis. The viewing system is also modeled as a pinhole system, with the principal ray parallel to the Z axis. The principal plane of the projector is the same as the viewing system. If the principle plane of the viewing and projection system are not the same, then there are additional nonlinear effects which we will not deal with here[3]. At a distance H from the principal plane of the projector and viewing system is the plane $Z \equiv \phi \equiv 0$. The pitch of the projected grating at $Z = 0$ is Po. By similar triangles, the pitch of the grating at other points is a function of Z, such that:

$$P = Po * \frac{(H - Z)}{H} \tag{1}$$

The phase is shifted by translating the projector in the X direction an amount Tx. The amount of phase shift, or ß, for a given translation "Tx" is:

$$\frac{\beta}{2\pi} = \frac{Tx}{P} = \frac{Tx}{Po} * \frac{H}{(H - Z)} \tag{2}$$

From figure 1, it can be seen that the relationship between Z and ϕ can be expressed as:

$$Z = \frac{\phi}{2\pi} P \frac{H}{d} = \frac{\phi}{2\pi} Po \frac{(H - Z)}{d} \tag{3}$$

Solving for Z in equation (3), we get:

$$Z = \frac{H \frac{\phi}{2\pi} * \frac{Po}{d}}{\left[1 + \frac{\phi}{2\pi} * \frac{Po}{d}\right]} \tag{4}$$

and solving for ϕ we get:

$$\frac{\phi}{2\pi} = \frac{d}{Po} * \frac{Z}{(H - Z)} \tag{5}$$

When the projector system is translated Tx, the distance d is changed a like amount. This results in a change of phase, as defined by equation (5). The actual intensity values for the six images is:

$$I_1 = I * (1 + c * \cos(\phi_1 - 2\beta)) \tag{6a}$$

$$I_2 = I * (1 + c * \cos(\phi_2 - \beta)) \tag{6b}$$

$$I_3 = I * (1 + c * \cos(\phi_3) \tag{6c}$$

$$I_4 = I * (1 + c * \cos(\phi_4 + \beta)) \tag{6d}$$

$$I_5 = I * (1 + c * \cos(\phi_5 + 2\beta)) \tag{6e}$$

$$I_6 = I * (1 + c * \cos(\phi_6 + 3\beta)) \tag{6f}$$

Where:

$$\phi_1 - 2\beta = 2\pi \frac{(d - 2Tx)}{Po} * \frac{Z}{(H - Z)} - 2\beta \tag{7}$$

$$\phi_2 - \beta = 2\pi \frac{(d - Tx)}{Po} * \frac{Z}{(H - Z)} - \beta \tag{8}$$

$$\phi_3 = 2\pi \frac{d}{Po} * \frac{Z}{(H - Z)} \tag{9}$$

$$\phi_4 + \beta = 2\pi \frac{(d + Tx)}{Po} * \frac{Z}{(H - Z)} + \beta \tag{10}$$

$$\phi_5 + 2\beta = 2\pi \frac{(d + 2Tx)}{Po} * \frac{Z}{(H - Z)} + 2\beta \tag{11}$$

$$\phi_6 + 3\beta = 2\pi \frac{(d + 3Tx)}{Po} * \frac{Z}{(H - Z)} + 3\beta \tag{12}$$

Since the change in ϕ with respect to the phase shift is proportional to the phase shift ß, we can redefine ϕ and ß to move the varying term from ϕ to ß. First we must separate the constant term related to the period from the changing term containing Tx:

$$\phi_i = 2\pi \frac{(d + (i-3)Tx)}{Po} * \frac{Z}{(H - Z)} \tag{13}$$

$$\phi_i = 2\pi \frac{d}{Po} * \frac{Z}{(H - Z)} + 2\pi \frac{(i-3)Tx}{Po} * \frac{Z}{(H - Z)} \tag{14}$$

After redefining ϕ and ß, we get:

$$\phi = 2\pi \frac{d}{Po} * \frac{Z}{(H - Z)} \tag{15}$$

And:

$$\beta = 2\pi \frac{Tx}{Po} * \frac{H}{(H - Z)} + 2\pi \frac{Tx}{Po} * \frac{Z}{(H - Z)} \tag{16}$$

Which reduces to:

$$\beta = 2\pi \frac{Tx}{Po} * \frac{(H + Z)}{(H - Z)} \tag{17}$$

Rewriting equation (17) in terms of Z yields:

$$Z = H \frac{\left[Po * \frac{\beta}{2\pi} - Tx \right]}{\left[Po * \frac{\beta}{2\pi} + Tx \right]} \tag{18}$$

Now using equations (4) and (18) we can state the relationship between ϕ and ß as:

$$H \frac{\left[Po \dfrac{\beta}{2\pi} - Tx \right]}{\left[Po \dfrac{\beta}{2\pi} + Tx \right]} = H \frac{\dfrac{\phi}{2\pi} * \dfrac{Po}{d}}{\left[1 + \dfrac{\phi}{2\pi} * \dfrac{Po}{d} \right]} \qquad (19)$$

Solving for ϕ, we get:

$$\phi = d \frac{(Po\ \beta - 2\pi Tx)}{2\ Po\ Tx} \qquad (20)$$

We rewrite the ϕ as:

$$\phi = 2\pi N + \phi' \qquad (21)$$

Where N is the unknown integer portion of the phase and ϕ' is the fractional, or modulo 2π part. Substituting the expression for ϕ in equation (20), and solving for N we get the exact solution of the relationship of ϕ', ß and N:

$$2\pi N = d * \frac{(Po\ \beta - 2\pi Tx)}{2\ Po\ Tx} - \phi' \qquad (22)$$

4. EXPERIMENTAL SIMULATION

To test the performance of the equations, we used a computer simulation. To simulate 8 bit grey scale, the range of intensity was set to an integer limited to the range of 0 - 256. To further simulate 8 bit grey scale, 2 bits of random noise was added to the signal. The data presented here was generated using the following values:

```
d  = 50 mm
H  = 400 mm
Po = 1 mm
Tx = .20 mm
c  = .5
I  = 128
```

The contour interval of the simulation at the plane Z = 0 is:

$$CI_o = Po * \frac{H}{d} = 1 * 400/50 = 8 \text{ mm} \tag{23}$$

Values of ß and ϕ relating to the Z map were generated via equations (2) and (7-11). Intensity values were generated via equations (6a-6f). The equation to extract ϕ' is:

$$\tan(\phi) = \frac{2*(I_2 - I_4)}{(2*I_3 - I_5 - I1)} \tag{24}$$

This commonly referred to as the 5 bucket algorithm because it uses 5 images to determine ϕ[5]. This algorithm was chosen because it is not sensitive to small changes in beta. Two different equations are used to calculate ß, based on ϕ. The two equations are based on the 5 bucket algorithm[5], with the first one using the first 5 buckets, and the second(shifted) using buckets 2-6.

$$\cos(\beta) = \frac{I_5 - I_1}{2*(I_4 - I_2)} \tag{25}$$

$$\cos(\beta) = \frac{I_6 - I_2}{2*(I_5 - I_3)} \tag{26}$$

The use of 6 images, with a shift of one image for the two equations, is necessary because the denominator in equation (25) goes to zero under certain conditions. The proper equation to use, based on ϕ, to minimize error was determined experimentally. Figure 2 shows the comparison of the maximum error of the two different algorithms for 5000 random inputs of ß and ϕ. The range of ß is 85 to 95 degrees, the range of ϕ is 0 to 360 degrees. To simulate actual camera input, equations (6a)-(6f) were used to generate intensity values. The contrast was set at 0.5, the intensity values were rounded, and 2 bits of random noise was added as noted earlier. Based on the graph, the rule in figure 3 was developed.

The output of the simulation is shown in Figures 4-12. Figure 4 is the profile of the original surface. This test surface contains extremely steep slopes and discontinuities that would normally be a problem to analyze. Figure 5 is the intensity image of bucket I_0 and I_4 for the interferogram generated from equations (6a)-(6f). As can be seen from the figure, there is a change in phase shift or ß with height in Z. Where the two curves overlap, ß = 90 degrees. Figure 6 is the fractional phase map ϕ' of the surface. This map was constructed using equation (24). Notice the 2π ambiguities. Figure 7 is the ß map of the surface. This ß map is somewhat noisy, but follows the shape of the part, with no gaps or discontinuities. Figure 8 is the fringe order number N, obtained from equation (22). Figure 9 is the integer fringe order number N after the rounding process. Figure 10 is the absolute phase map of the surface. This is the result of the combination of the integer fringe order number N and the phase map ϕ' per equation (21). Figure 11 is the resultant Z map of the surface from equation (3). Figure 12 is the deviation from the original Z map of the model.

5. DISCUSSION

Equation (22) predicts the fringe order number fairly well although it is very noisy. The reason it works so well can be found in the round off function, where N is converted from a real number to an integer. If ß is too noisy, which occurs at lower contrast and smaller contour intervals, then the rounding process may produce an error. There are several possible solutions to this problem. One can run some sort of smoothing filter to reduce the noise on the signal. One can also increase the ratio of the change in ß verse the fringe order. The derivative of the ß with respect to ϕ is:

$$\frac{d\beta}{d\phi} = \frac{Tx}{d} \qquad (27)$$

This relationship between ß and ϕ implies that an increase in the shift from 90 degrees to 360 + 90 degrees would reduce the error. This has the effect of increasing the change in ß per fringe order. There is a problem however, with too great a change in ß, because the five bucket algorithm has an increasing error in ϕ as ß moves farther from 90 degrees. Using a 120 degree shift algorithm and changing the nominal ß from 90 to 120 degrees would also help. The problem associated with too great a change in ß can be solved by using the four bucket ß compensated algorithm[4] which is less sensitive to changes in ß and has a minimum error at ß = 120 degrees. Since Tx is a function of Po, (Tx = .25*Po for 90 degree shift), we can rewrite the derivative of ß with respect to ϕ:

$$\frac{d\beta}{d\phi} = \frac{Po}{d} \qquad (28)$$

This implies that a system with a large contour interval will perform better than a more sensitive system with a small contour interval for a given standoff distance H.

In a real moire setup, when the projector is shifted, the point on the model surface will receive light from a different part of the light source. This change in lighting must be minimized, as the algorithms are sensitive to changes in the background lighting and fringe contrast, if they occur between exposures.

The ß algorithms are also sensitive to the true shape of the fringes. For the best results, the fringes must be of a sinusoidal nature. Problems may occur if the fringes are formed from square wave gratings, or the camera that records the image does not have a highly linear response.

When implementing the field shift setup, one must know the pitch of the grating, Po at Z = 0. This can be determined by solving equation (2) for Po at Z = 0 and measuring ß :

$$Po = \frac{2\pi}{\beta} Tx \qquad (29)$$

One can run into problems if Po is not constant, but rather changes over the field due to the keystone affect, barrel distortion, and other aberrations . These errors can be compensated for by solving for and storing the value of Po for every pixel. Such a pixel map of Po can be made during a system calibration by placing a flat plane at the Z = 0 plane, measuring ß at every point, and calculating Po for every pixel.

This technique may also be extended to other forms of interferometry, provided a means of shifting the field and a dependence of ß on the function being measured can be established. One side benefit of the field shifting algorithm is that it is an inherently parallel process. This indicates that extremely high computation rates can be achieved by implementing the algorithm on a parallel processor.

6. CONCLUSION

A new technique is presented to produce absolute phase and Z (depth) information from a moire system. This technique may also be applicable to other forms of phase shift interferometry for such possible applications as absolute analysis of temperature, pressure, deformation, and thickness using well known, non-invasive, interferometric techniques. Simulated data using the field shift technique showed usable results in contouring a difficult to analyze object which included discontinuous jumps in the surface. Potential limitations of this technique in practice would include non-sinusoidal fringe data and low fringe contrast. A strength of field shifting in practice will be its applicability to fast, parallel process computers, thereby allowing for fast absolute contour generation of prismatic parts[6].

7. REFERENCES

1. K. Creath, "Step height measurement using two-wavelength phase shifting interferometry", Appl. Opt., 26, 2810 (1987).

2. G.T. Reid, R.C. Rixion, S.J. Marshall, "3-D machine vision for automatic measurements of complex shapes", (probably:) Optics and Lasers in Eng., Vol. 6 (1985).

3. Harding, K.H., Boehnlein, A.J., Michniewicz, M. A., Small Angle Moire Contouring, Proceedings of 1987 Conference on Advances in Intelligent Robotics Systems, SPIE, Cambridge, Massachusetts, November 1-6, 1987, Proceedings No. 850 (1987).

4. Y.Y. Cheng, J.C. Wyant, "Phase shifter calibration in phase-shifting interferometry", Appl. Opt. 24 3049 (1985)

5. P. Hariharan, B.F. Oreb, and T. Eiju, "Digital phase-shifting interferometry: a simple error-compensating phase calculation algorithm", App. Opt. 26, (1987)

6. Albert J. Boehnlein and Kevin G. Harding, "Adaption of a parallel architecture computer to phase shifted moire interferometry," Optics, Illumination and Image Sensing for Machine Vision, SPIE, Cambridge, MA Proceesings No. 728, p. 183 (1986).

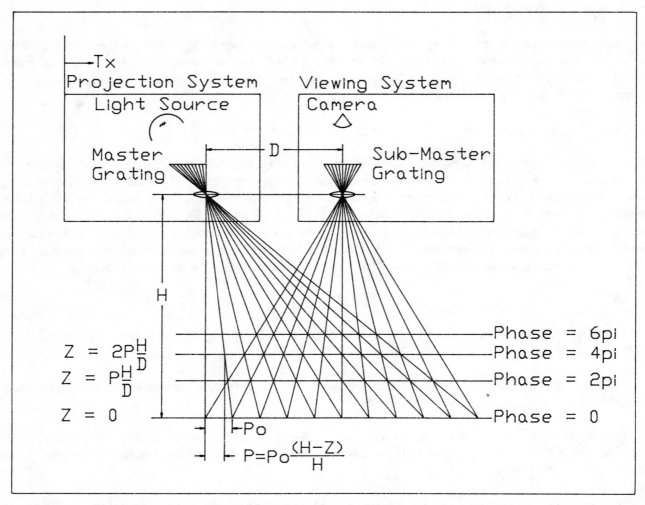

Figure 1, Diagram of field shift moire system.

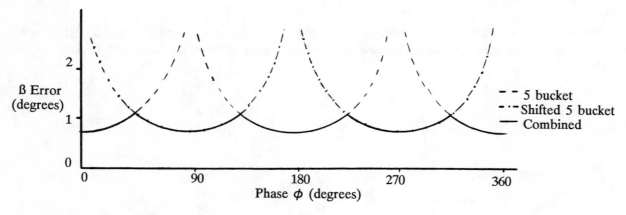

Figure 2,
Plot of error in β algorithm versus ϕ

ϕ (degrees)	β algorithm
0-45	shifted 5 bucket
45-135	5 bucket
135-225	shifted 5 bucket
225-315	5 bucket
315-360	shifted 5 bucket

Figure 3, Table of which β algorithm to use based on ϕ.

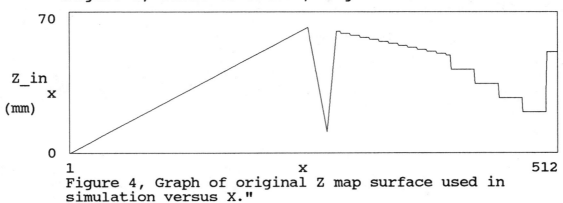

Figure 4, Graph of original Z map surface used in simulation versus X."

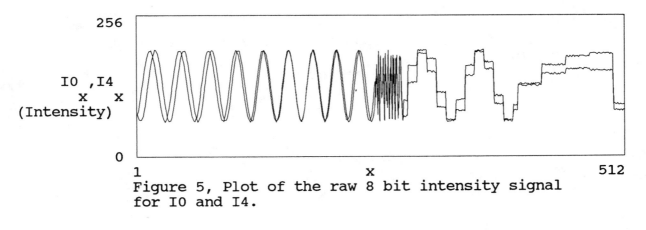

Figure 5, Plot of the raw 8 bit intensity signal for I0 and I4.

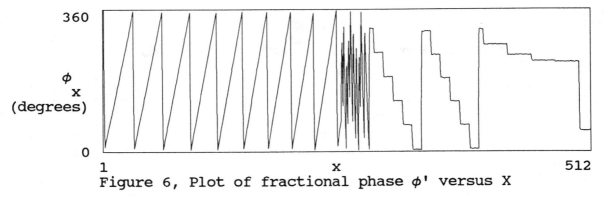

Figure 6, Plot of fractional phase ϕ' versus X

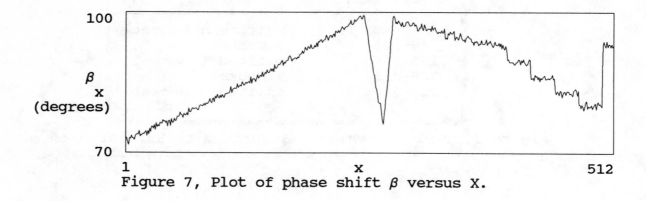

Figure 7, Plot of phase shift β versus X.

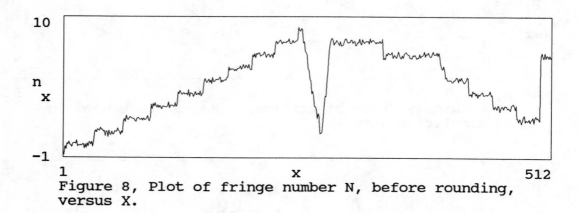

Figure 8, Plot of fringe number N, before rounding, versus X.

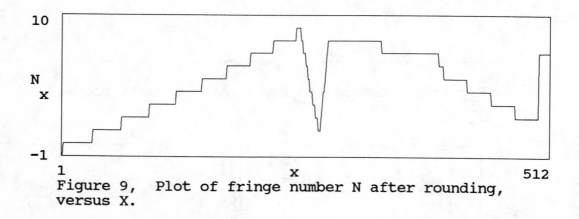

Figure 9, Plot of fringe number N after rounding, versus X.

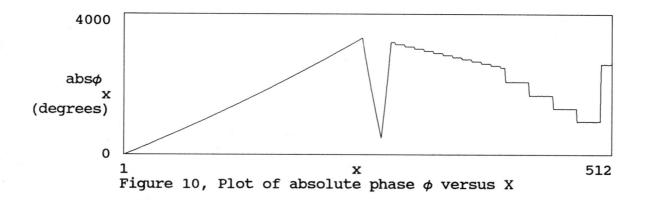

Figure 10, Plot of absolute phase φ versus X

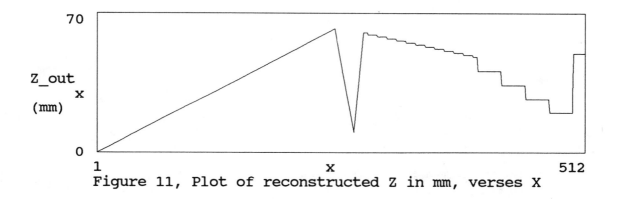

Figure 11, Plot of reconstructed Z in mm, verses X

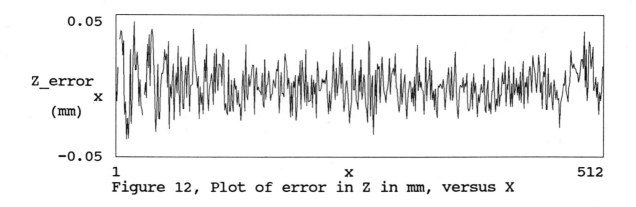

Figure 12, Plot of error in Z in mm, versus X

Sinusoidal phase modulating laser diode interferometer with feedback control system to eliminate external disturbance

Osami Sasaki, Kazuhide Takahashi, and Takamasa Suzuki

Niigata University, Faculty of Engineering
8050 Ikarashi 2, Niigata-shi, Japan

1. ABSTRACT

We propose a sinusoidal phase modulating laser diode interferometer which is insensitivity to vibrations of optical devices and fluctuations in the optical wavelength of the laser diode. We analyze the sinusoidal phase modulation in a laser diode interferometer, and describe the principle of the feedback control of the injection current of the laser diode to eliminate the fluctuations in the phase of the interference signal caused by external disturbances. We construct two sinusoidal phase modulating interferometers for movement measurements and surface profile measurements, respectively. The experimental results make it clear that the interferometers can be used in mechanically noisy circumstances.

2. INTRODUCTION

Heterodyne interferometry and fringe scanning interferometry have been widely used to measure surface profiles with high accuracy. Recently laser diodes (LDs) have been incorporated into heterodyne interferometers[1] and fringe scanning interferometers[2] as light sources and phase modulators. As another interferometric technique, we proposed sinusoidal phase modulating (SPM) interferometry, in which surface profiles are obtained with the Fourier transform method[3] and the integrating-bucket method.[4] We also reported the method of movement measurements in SPM interferometry.[5]

In this paper we describe a SPM laser diode interferometer which is insensitive to vibrations of optical devices and fluctuations in the optical wavelength of the LD. Sinusoidal phase modulated interference signal is generated by modulating the injection current of the LD with a sinusoidal wave signal. The phase modulation in SPM interferometry is very simple compared with those in heterodyne interferometry and fringe scanning interferometry. The signal that is a trigonometric function of the phase difference between the object and reference waves can be easily obtained from the sinusoidal phase modulated interference signal. This signal is used as a feedback signal in controlling the injection current of the LD to reduce the fluctuations in the phase of the interference signal caused by external disturbances. These special characteristics of the sinusoidal phase modulated interference signal allow us to construct an interferometer with the feedback control system to eliminate the external disturbances.

The sinusoidal phase modulation in a LD interferometer is theoretically analyzed in Sec.2, and the principle of the feedback control of the injection current is described in Sec.3. In Sec.4, we describe a SPM interferometer for movement measurements in which we obtain a feedback signal from the interference signal generated with a stationary object. We measure movements of a piezoelectric transducer without suffering from external disturbances. In Sec.5, we also constrict a SPM interferometer for surface profile measurements, and measure surface profiles of diamond-turned aluminum disks. The measurement repeatability is greatly improved by the feedback control of the injection current. The experimental results makes it clear that the SPM interferometers presented here can be used in mechanically noisy circumstances.

3. SINUSOIDAL PHASE MODULATION IN LASER DIODE INTERFEROMETER

Let us consider a Twyman-Green type interferometer as shown in Fig.1. The injection current of a LD consists of a DC component i_0 and a time-varying component $\Delta i_c(t)$ as follows:

$$i(t) = i_0 + \Delta i_c(t) . \tag{1}$$

The DC component determines a central wavelength of the light λ_0, and the Δi_c produces a small change in the wavelenght of the LD

$$\Delta\lambda(t) = \beta\Delta i_c(t) . \tag{2}$$

Then the wavelength of the LD is given by

$$\lambda(t) = \lambda_0 + \Delta\lambda(t) . \tag{3}$$

The optical wave emitted from the LD is represented by

$$\exp\{j2\pi c \int_0^t [1/\lambda(t)]dt\} = \exp\{j\phi(t)\} , \tag{4}$$

where c is the velocity of the light. The light reflected from an object is an objective wave, and the light reflected from a mirror (M) is a reference wave. The optical path length of these waves are denoted by l_0 and l_r, respectively. The objective wave U_0 and reference wave U_r on the photodiode (PD) are represented by

$$U_0 = \exp[j\phi(t - \tau_0)], \qquad U_r = \exp[j\phi(t - \tau_r)] , \tag{5}$$

where $\tau_0 = l_0/c$ and $\tau_r = l_r/c$. The time-varying component of the interference signal produced with the two waves is given by

$$S(t) = \cos[\phi(t - \tau_0) - \phi(t - \tau_r)] = \cos\Phi(t) . \tag{6}$$

Using the approximation

$$1/\lambda(t) \simeq (1/\lambda_0)\{1 - [\Delta\lambda(t)/\lambda_0]\} , \tag{7}$$

and the definition

$$\int\Delta\lambda(t)dt = \Delta\Lambda(t) , \tag{8}$$

the argument of Eq.(6) becomes

$$\Phi = (2\pi/\lambda_0)l - (2\pi c/\lambda_0^2)[\Delta\Lambda(t - \tau_0) - \Delta\Lambda(t - \tau_r)] , \tag{9}$$

where $l = l_r - l_0$. In the condition of $\tau_r - \tau_0 \ll 1$, we have the approximation

$$\Delta\Lambda(t - \tau_0) - \Delta\Lambda(t - \tau_r) \simeq (1/c)\Delta\lambda(t) . \tag{10}$$

For sinusoidal phase modulation,

$$\Delta i_c(t) = a \cos(\omega_c t + \theta) , \tag{11}$$

we obtain the interference signal

$$S(t) = \cos[z \cos(\omega_c t + \theta) + \alpha] , \tag{12}$$

where

$$z = -(2\pi/\lambda_0^2)\beta al, \qquad \alpha = (2\pi/\lambda_0)l . \tag{13}$$

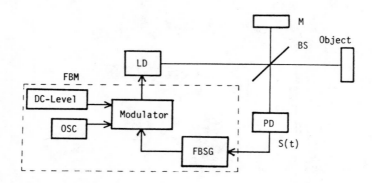

Fig. 1. Feedback control system in sinusoidal phase modulating laser diode interferometer.

4. ELIMINATIO OF EXTERNAL DISTURBANCES WITH FEEDBACK CONTROL

The wavelength of the LD changes by $\Delta\lambda_T$ with temperature. Optical devices in the interferometer vibrate in response to external mechanical vibrations. This caused the change Δl in the optical path difference between the object and reference waves. These $\Delta\lambda_T$ and Δl cause the fluctuation in the phase of the interference signal. The fluctuation is compensated by controlling the injection current to produce the change $\Delta\lambda_I$ in the wavelength of the LD. Considering the changes $\Delta\lambda_T$, Δl and $\Delta\lambda_I$ in Eqs.(9) and (10), l is replaced with $l + \Delta l$ and $\Delta\lambda$ is taken to be $\Delta\lambda + \Delta\lambda_T + \Delta\lambda_I$. By neglecting the term of $(\Delta\lambda + \Delta\lambda_T + \Delta\lambda_I)\Delta l$, the interference signal is written as

$$S(t) = \cos[z \cos(\omega_c t + \theta) + \alpha + \delta(t)] , \qquad (14)$$

where

$$\delta(t) = (2\pi/\lambda_0)\Delta l - (2\pi l/\lambda_0^2)(\Delta\lambda_T + \Delta\lambda_I) . \qquad (15)$$

We try to reduce the phase $\delta(t)$ to zero by controlling the injection current or $\Delta\lambda_I$.

Let us explain how to generate the feedback signal for the injection current of the LD. The expansion of Eq.(14) is given by

$$S(t) = \cos[\alpha + \delta(t)][J_0(z) - 2J_2(z)\cos(2\omega_c t + 2\theta) + \cdots]$$
$$-\sin[\alpha + \delta(t)][2J_1(z)\cos(\omega_c t + \theta) - 2J_3(z)\cos(3\omega_c t + 3\theta) + \cdots] . \qquad (16)$$

Producing the signal $S(t)\cos(\omega_c t + \theta)$ and passing this signal through a low-pass filter, we obtain the following feedback signal which is the output of the feedback signal generator (FBSG) shown in Fig.1:

$$J_1(z)\sin[\alpha + \delta(t)] . \qquad (17)$$

This feedback signal is available in the region of $z = 0.5 - 3.5$ where the value of the $J_1(z)$ is not so small. When the phase α is nearly multiplies of π rad, we can keep the phase $\delta(t)$ to be zero stably with a proportional feedback control using the feedback signal given by Eq.(17). The phase α is adjusted with the DC component of the injection current. The portion blocked with dot lines in Fig.1 is referred to feedback modulator (FBM). This modulator produces the injection current of the LD which is controlled so that the phase $\delta(t)$ is reduced to zero.

5. MOVEMENT MEASUREMENTS

5.1. Interferometer

Figure 2 shows a SPM interferometer with feedback control system for movement measurements. The light emitted from a LD is collimated with a lens 1 (L1). The light reflected from a mirror 1 (M1) is a reference wave. The light passed through a beam splitter (BS) is an object wave. A portion of the object wave is illuminated onto an object through a lens 2 (L2). The movement of the object is represented by $r(t)$. The reflected light from the object and the reference light are superimposed on a photodiode 1 (PD1). The interference signal detected with the PD1 is written as

$$S_1(t) = \cos[z_1 \cos(\omega_c t + \theta) + \alpha_{10} + \alpha_1(t) + \delta_1(t)] , \qquad (18)$$

where $\alpha_1(t) = (4\pi/\lambda_0)r(t)$, and the α_{10} is a constant. On the other hand, the rest of the object wave is illuminated onto a mirror 2 (M2). The reflected light and the reference light are deflected with a prism, and reach to a photodiode 2 (PD2). The interference signal detected with PD2 is written as

$$S_2(t) = \cos[z_2 \cos(\omega_c t + \theta) + \alpha_{20} + \delta_2(t)] . \qquad (19)$$

The feedback signal is generated from this interference signal in the FBM. When the feedback control operates well, the phase δ_2 is reduced to a small value $\Delta\delta$ as follows:

$$\delta_2(t) = (2\pi/\lambda_0)\Delta l_2 - (2\pi l_2/\lambda_0^2)(\Delta\lambda_T + \Delta\lambda_I) = \Delta\delta . \qquad (20)$$

Then, the phase $\delta_1(t)$ is written as

$$\delta_1(t) = (2\pi/\lambda_0)[\Delta l_1 - (l_1/l_2)\Delta l_2] + (l_1/l_2)\Delta\delta . \qquad (21)$$

Since the optical path length l_1 is longer than the optical path length l_2 and the change Δl_1 is not completely equal to the change Δl_2 in this interferometer, the phase fluctuation $\delta_1(t)$ cannot be always reduced to the amount $\Delta\delta$.

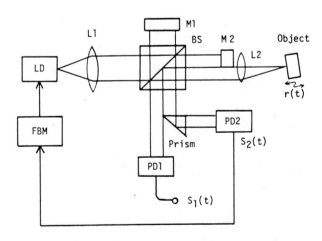

Fig. 2. SPM laser diode interferometer with feedback control system for movement measurements.

5.2. Experimental Results

We measured movements of the piezoelectric transducer which vibrated sinusoidally with a frequency of 100 Hz. The frequency of the sinusoidal phase modulation was 1KHz and the cutoff frequency of the low-pass filter employed in the FBM was 200 Hz. The movement $r(t)$ and the phase $\delta_2(t)$ were obtained by using the method described in Ref.5. Figures 3 and 4 show the movement $r(t)$ and the phase $\delta_2(t)$ measured when the feedback control did not operate. The measured movement contains the phase fluctuation $\delta_1(t)$ which is almost equal to the measured phase $\delta_2(t)$. Figures 5 and 6 show the movement $r(t)$ and the phase $\delta_2(t)$ measured when the feedback control operated well. The measured $\delta_2(t)$ corresponds to the $\Delta\delta$ in Eq.(20). The optical path lengths l_1 and l_2 were 20 mm and 15 mm, respectively. The phase $\delta_1(t)$ is reduced to be a small value, and the measured movement can be regarded to be a sinusoidal wave.

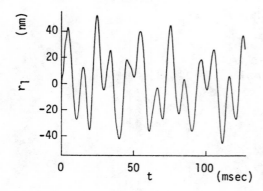

Fig. 3. Movement $r(t)$ measured when the feedback control did not operate.

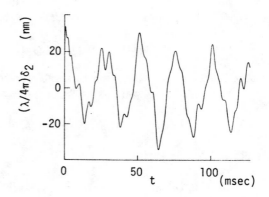

Fig. 4. Phase $\delta_2(t)$ measured when the feedback control did not operate.

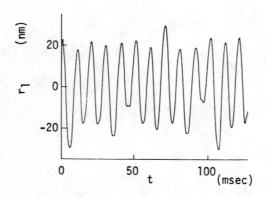

Fig. 5. Movement $r(t)$ measured when the feedback control operated.

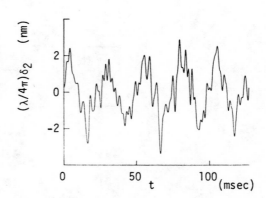

Fig. 6. Phase $\delta_2(t)$ measured when the feedback control operated.

6. SURFACE PROFILE MEASUREMENTS

6.1. Interferometer

Figure 7 shows a SPM interferometer with feedback control system for surface profile measurements. The lens 2 (L2) makes an image of an object on a linear CCD image sensor. The surface profile of the object is represented by $r(x)$. The light reflected from the mirror 1 (M1) is a reference wave. The light near the CCD image sensor is reflected by the mirror 2 (M2) and reaches to the photodiode (PD). The interference signal detected with the CCD image sensor is written as

$$S_1(t,x) = \cos[z\cos(\omega_c t + \theta) + \alpha_{10} + \alpha_1(x) + \delta(t)] , \qquad (22)$$

where $\alpha_1(x) = (4\pi/\lambda_0)r(x)$. The interference signal detected with the PD is written as

$$S_2(t) = \cos[z\cos(\omega_c t + \theta) + \alpha_{20} + \delta(t)] . \qquad (23)$$

Since the distance between the measuring points for the CCD image sensor and the PD is short, the phase fluctuations in the signals $S_1(t)$ and $S_2(t)$ are considered to be identical. In other words, the conditions of $l_1 = l_2$ and $\Delta l_1 = \Delta l_2$ hold in this interferometer. The signal $S_2(t)$ is fed to the FBM to generate the feedback signal. This feedback control system reduce the phase fluctuation $\delta(t)$ in the signal $S_1(t)$ to $\Delta\delta$.

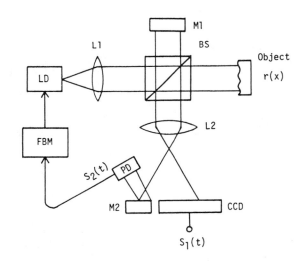

Fig. 7. SPM laser diode interferometer with feedback control system for surface profile measurements.

6.2 Experimental Results

We measured surface profiles of diamond-turned aluminum disks. The surface profile was obtained from the CCD output using the Fourier transform method described in Ref. 3. The same surface profile was measured at an interval of a few minutes. Figure 8 shows two surface profiles measured at the interval when the feedback control did not operate. There are slight differences between the two surface profiles. The measurement repeatability was between about 3.5 nm and 7.0 nm. Figure 9 shows two surface profiles measured at the interval when the feedback control operated well. The two surface profiles are almost identical. The measurement repeatability was greatly improved by the feedback control and was between about 0.5 nm and 1.0 nm.

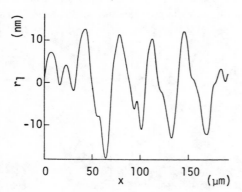 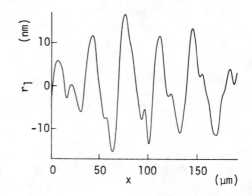

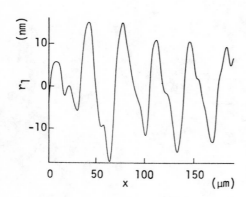 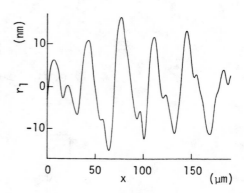

Fig. 8. Two surface profiles measured at an interval of a few minutes when the feedback control did not operate.

Fig. 9. Two surface profiles measured at an interval of a few minutes when the feedback control operated.

7. CONCLUSIONS

We constructed the SPM LD interferometers which were very insensitivity to external disturbances such as mechanical vibrations and fluctuations in temperature. Since the optical path lengths and its fluctuations in the two different interference signals are not equal in the interferometer for the movement measurements, the measured movements contain phase fluctuations which can not be eliminated by the feedback control. However, the measured movements approach to the real movements. On the other hand, since they are equal in the interferometer for the surface profile measurement, the phase fluctuations are almost eliminated to improve greatly the measurement repeatability. The experimental results show clearly that the SPM LD interferometers presented in this paper can be used in mechanically noisy circumstances.

8. REFERENCES

1. K. Tatsuno and Y. Tsunoda, "Diode Laser Direct Modulation Heterodyne Interferometer," Appl. Opt. 26, 37 (1987).

2. Y. Ishii, J.Chen, and K.Murata, "Digital Phase-Measuring Interferometry with a Tunable Laser Diode, " Opt. Lett. 12, 233 (1987).

3. O. Sasaki and H.Okazaki, "Sinusoidal Phase Modulating Interferometry for Surface Profile Measurement, " Appl. Opt. 25, 3137 (1986).

4. O.Sasaki , H.Okazaki, and M.Sakai, " Sinusoidal Phase Modulating Interferometer Using the Integrating-Bucket Method," Appl. Opt. 26, 1089 (1987).

5. O.Sasaki and K.Takahashi, "Sinusoidal Phase Modulating Interferometer Using Optical Fibers for Displacement Measurement," Appl. Opt. 27, 4139 (1988).

Non-contact ranging using dynamic fringe projection

M.M. Shaw, D.M. Harvey, C.A. Hobson, M.J. Lalor

Coherent and Electro-Optics Research Group
Liverpool Polytechnic, Byrom Street
Liverpool, England, L3 3AF

ABSTRACT

Major advances in imaging technology have seen the advent of 3-dimensional imaging systems, and the ensuing advantages that they have over 2-dimensional systems in many situations. Both passive and active systems for 3-dimensional image production have been widely used for robot control, and for high accuracy measurements in engineering metrology.

This paper will concentrate on a discussion of the development of a new instrument employing dynamic fringe projection techniques, capable of simultaneously measuring the range at each point in the field of view of, say, a CCD camera in real time. The instrument, called the Dynamic Automated Range Transducer, (DART), uses a method of fringe projection based upon the principle of triangulation. The current investigations aimed at automating the ranging process will be described, and an error analysis and theoretical maximum resolution of the system will be included.

1. INTRODUCTION

Inspection of industrial components is becoming increasingly important for the production of high quality products, and components that will easily integrate together in a systems environment. Equally important in an automated assembly or manufacturing process, is the control of .such a production facility so that the required component or system specification is achieved. Such inspection and control systems may benefit greatly by the application of integrated vision systems.

Conventional two-dimensional image processing systems have been used for many years, and in very many cases are wholly satisfactory in terms of the usefulness of the information that they yield. For example, a simple image subtraction process will show the presence or otherwise, of a component on a printed circuit board. A system previously described[1], uses the information contained in a subtraction image, generated using a fringe projection technique, from images showing components before wear and after wear, to evaluate the volume of a wear scar. Of course, a subtraction process such as this, requires accurate alignment of the component under test, with what is effectively the reference image. A method of automatically aligning such images using correlation techniques, based upon digital signal processing (DSP) modules, is currently being investigated[2]. The systems described above work on two-dimensional data. The third dimension can readily be computed, increasing the range of applications that such a system inevitably generates. The 3-dimensional examination of wear scars would allow the shape of the wear scar to be evaluated yielding information about the action of the wearing process.

2. DYNAMIC FRINGE PROJECTION

The techniques previously outlined for wear scar evaluation are based on the shadow projection of a Ronchi grating, and may be termed static fringe projection systems. The resulting deformation of the contours caused by the surface of the object, yields information about the surface. However, the height information describing it is relative in nature. The height of a particular point can thus only be described by reference to other points on the surface. Furthermore, the height information may be ambiguous in terms of whether the surface is concave or convex. Phase stepping techniques based either on holographic or Moiré systems, can solve the height ambiguity problem[3].

An alternative arrangement which will yield unambiguous height information of an absolute nature, is one which analyses a rotating fringe pattern. This is termed dynamic fringe projection, and is the technique exploited in the Dynamic Automated Range Transducer (DART), currently being developed. Fig.1 shows a block schematic diagram of the system. A point source of light is produced using a laser and spatial filter arrangement. The point source is projected through a rotating cosinusoidal transmission grating, and the dynamic fringe pattern is imaged along the optical axis onto the detector circuits. The range of the target is related to the number of fringes which pass over it as the grating rotates through a known angle. The grating is rotated using a computer controlled stepping motor and a gearing arrangement located around the outside of the grating itself. An optical shaft encoder together with decoding circuits, provide grating position information to the processor.

Fig.2 shows the optical arrangement and geometry in more detail. The grating is a holographic recording of Young's fringes on a glass plate generated using a simple double pinhole interferometer arrangement. The pitch of the grating, Po, is of the order of 1mm and is kept deliberately large to avoid diffraction effects. By a process of similar triangles and by taking into account the rotation of the grating, it has been shown[4] that the range of a single point located on the optical axis of the system is given by

$$Z = \frac{ZfNPo}{YsSin\theta - NPo} \tag{1}$$

where Po is the pitch of the grating
    N is the fringe order
    θ is the grating angle at which that value of N is measured
    Ys is the perpendicular distance between source and detector
    Zf is the distance between grating and detector along the optical axis

However, for points not located on the optical axis, the range equation will differ depending upon the distance from the centre of the detector that a particular pixel images a point on the target. This is illustrated in fig.3, and the equations for the range of the two points shown are

$$Z1 = \frac{ZfN1Po}{(Ys+X)Sin\theta - N1Po} \tag{2}$$

$$Z2 = \frac{ZfN2Po}{(Ys-L)Sin\theta - N2Po} \tag{3}$$

where N1 and N2 are the fringe orders at points 1 and 2 respectively
     X and L are the distances of the respective pixels from the optical axis

Therefore, if the location of the detector is accurately known and the fringe order can be determined from the optical signal at each pixel, then the range of each point in the measurement space imaged by the detector, can be found. The x-y location of a pixel defines the location of the corresponding point on the target in the x-y plane. Analysis of the detected optical signal diffusely reflected from that point, produces the range or z co-ordinate information. Hence, a three-dimensional vision system may be realised.

3. ANALYSIS OF DETECTED SIGNALS

The amount of data generated from even a single point measurement system is quite large. This necessarily leads to large computer memory requirements and long processing times. The evaluation of this technique is, therefore, currently being carried out for a single point measurement located on the optical axis.

It has already been shown[4], that the form of the detected optical signal from a point on the surface of an object is

$$I(z) = \cos^4(N\pi Sin\theta) \tag{4}$$

Fig.4 shows an example of the type of waveform this yields where the fringe order N = 5.18. It can be seen that the number of peaks and fractions of a peak, as the grating rotates through 90 degrees, is the fringe order N. It is required to determine that value of N so that the range of the target can be found. The waveform shown in fig.4 is based on computer data simulations and contains an added Gaussian noise component to emulate a real signal.

A technique currently being investigated, is one which uses the differential of the received optical signal as a means of identifying the number of peaks or fringes. However, one must bear in mind the difficulties that the differential of any waveform containing noise, can produce. By judicious use of digital filtering techniques, these difficulties can be minimised. The derivative of the ideal waveform of equation (4), is

$$\frac{dI(z)}{d\theta} = -4N\pi\cos\theta\cos^3(N\pi sin\theta)sin(N\pi sin\theta) \tag{5}$$

When applied to noisy data such as that shown in fig.4, the resulting waveform of the differential after substantial filtering, is shown in fig.5. This shows the section of the differential waveform between 90 and 180 degrees rotation of the grating. By examination, the number of complete fringes can be found by counting the number of zero crossings and dividing by two. The angle θ, is given by the first zero crossing. As the waveform is symmetrical about $90°$ this gives the position of the last fringe peak. This method works satisfactorily if the fractional part of the fringe order is less than 0.5. If the fractional part of the fringe order is greater than 0.5, then the first zero crossing point is ignored. The second zero crossing point gives the required angle, and the total number of zero crossing points is again equal to twice the fringe order. This system may be implemented on a computer to simply evaluate the derivative signal in real-time, as the input data is received. It has the advantage that actual signal levels are not important, but only the number and position of zero crossing points need be considered. It has the disadvantage, however, of being very noise conscious and can introduce erroneous zero crossing positions. A number of digital filtering algorithms can be applied to the data as a means of minimising the occurrence of these errors. Filters implemented using software techniques, can introduce substantial processing times, which needs to be considered when assimilating a large amount of data. Hardware digital filters based on DSP's, would present a better solution as the filtering process could occur in real-time as the data arrived for processing.

4. RESULTS

The results shown in table 1 are based on data generated using a computer simulation of range measurements made on the optical axis of such a system. From the generated noisy data, filtering and derivative algorithms are used to count and locate the fringe peaks. This information is used to evaluate the range. The data generated by the simulation is based upon the following geometrical parameters

$$Zf = 100mm$$
$$Ys = 100mm$$
$$Po = 1mm$$

ACTUAL RANGE (mm)	SIGNAL/NOISE = S		SIGNAL/NOISE = 2S	
	MEASURED RANGE (mm)	% ERROR	MEASURED RANGE (mm)	% ERROR
10	12.489	+24.89	10.689	+6.89
20	20.702	+3.51	23.458	+17.29
30	29.991	-0.03	29.991	-0.03
40	45.003	+12.51	42.971	+7.43
50	49.983	-0.03	49.983	-0.03
60	63.951	+6.59	59.992	-0.01
70	69.982	-0.03	69.982	-0.03
80	79.976	-0.03	79.976	-0.03
90	86.316	-4.09	86.316	-4.09
100	95.929	-4.07	95.987	-4.01

TABLE 1. RESULTS USING SIMULATED DATA

Work is currently being undertaken to apply the knowledge gained using this simulation, to a real system. As far as possible, algorithms used for the filtering of data and generation of derivative information, will be implemented in hardware. This will allow the processing of the data in real-time which is particularly important when considering a multiple pixel measurement system. Attempts to identify and eliminate the cause of relatively large errors, especially at small ranges, will be made. However, it can be seen that accuracies in the order of 17 microns over 50mm, (99.97%), can be achieved with this system.

5. SYSTEM ERROR ANALYSIS AND RESOLUTION

To minimise system errors caused by the geometry itself and to examine the effects on the accuracy of the measured range, the range equation in terms of the total differential is analysed. The total differential is

$$dZ = \frac{Z(Z+Zf)}{Zf}\left[\frac{dN}{N} - \frac{d\theta}{\tan\theta} + \frac{dPo}{Po} - \frac{dYs}{Ys} + \frac{dZf}{Z+Zf}\right] \qquad (6)$$

By examination any errors in the range can be kept to a minimum by ensuring that both Zf and Ys are as large as possible. The resolution of the system will depend upon how accurately the fringe order N can be determined. Also it is assumed that errors in the measurement of the grating angle are minimal. The resolution may be defined as

$$dZ = \frac{Z(Z+Zf)dN}{ZfN} \qquad (7)$$

where dZ is the resolution
   dN is the smallest possible measured value of fringe order
Using the same system parameters generated for the previous data over a range of 50mm, and assuming that dN = 0.01, then dZ = 40 microns.

6. CONCLUSIONS

This paper has shown that the techniques of dynamic fringe projection are capable of producing a powerful system for the three-dimensional examination of industrial components. The work thus far has concentrated on the evaluation and development of a system capable of measuring the range of a single point located on the optical axis. However, it is also shown that by accurately knowing the position of each pixel in an array detector, the range at each point may be determined and a three-dimensional vision system is realised. An array detector such as a CCD camera would enable the range at each point in the field of view to be measured simultaneously. The processing times and memory requirements involved in such a system would be quite enormous. Work currently in progress is examining the use of pipelined/arrayed Transputer systems for high speed processing of such data. Dedicated DSP modules such as the Texas TMS-320 series devices, are likely to be most versatile for this type of application.

7. REFERENCES

1. M.B.Koukash, C.A.Hobson, M.J.Lalor, J.T.Atkinson, "Detection and Measurement of Surface Defects by Automatic Fringe Analysis", _Automatic Fringe Analysis_, Ed. B.L.Button, G.T.Reid, pp.97-107, FASIG, Loughborough, 1986.
2. D.A.Hartley, C.A.Hobson, S.Monaghan, "Correlation of Fringe Patterns Using Multiple Digital Signal Processors", _Fringe Pattern Analysis_, Proc.SPIE Vol.1163 San Diego, California, August 1989, to be published.
3. G.T.Reid, R.C.Rixon, H.I.Messer, "Absolute and Comparative Measurements of 3D Shape by Phase Measuring Moiré Topography", Optics and Laser Technology, Vol.16 No.6 pp.315-319, 1984.
4. M.M.Shaw, J.T.Atkinson, D.M.Harvey, C.A.Hobson, M.J.Lalor, "Range Measurement Using Dynamic Fringe Projection", J.of Physics D: Appl.Phys, Vol.21, pp.S4-S7, October 1988.

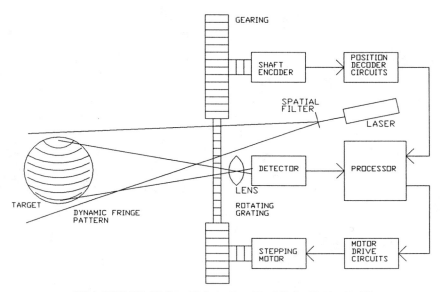

FIG. 1 SCHEMATIC DIAGRAM OF DYNAMIC AUTOMATED RANGE TRANSDUCER

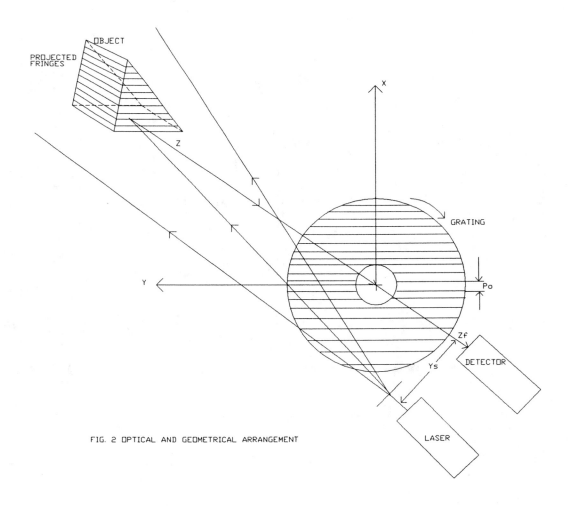

FIG. 2 OPTICAL AND GEOMETRICAL ARRANGEMENT

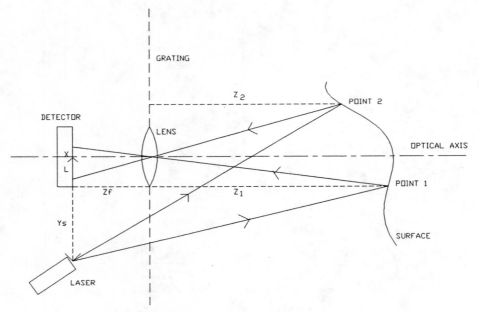

FIG. 3 GEOMETRY FOR POINTS NOT ON OPTICAL AXIS

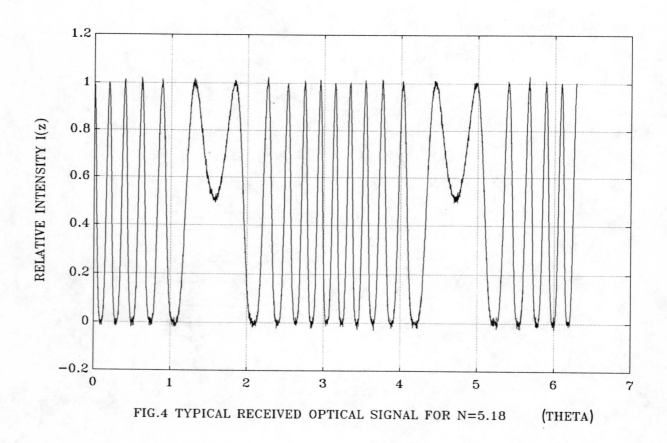

FIG.4 TYPICAL RECEIVED OPTICAL SIGNAL FOR N=5.18   (THETA)

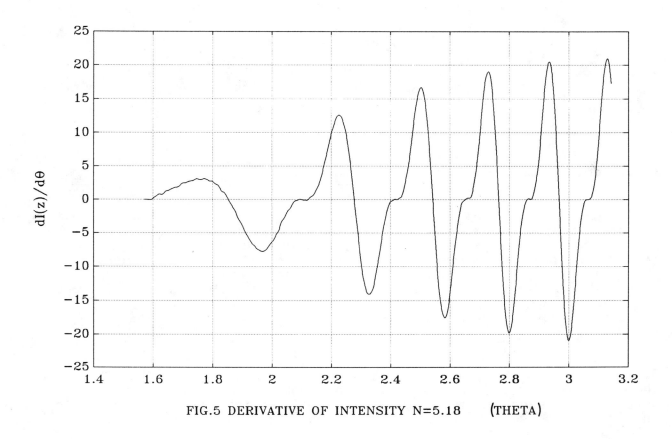

FIG.5 DERIVATIVE OF INTENSITY N=5.18   (THETA)

High-precise retardation measurement
Using a phase detection of Young's fringes

Suezou Nakadate

Tokyo Institute of Polytechnics, Department of Engineering
1583 Iiyama, Atsugi, Kanagawa 243-02, Japan

ABSTRACT

A method for a phase detection of Young's fringes is applied to a high-precise retardation measurement. A simple common-path polarizing interferometer is used with a birefringent wedge and a polarizer. The birefringent wedge introduces a spatially linear phase difference between orthogonally polarized light and Young's fringes are formed on an image sensor. The phase difference between orthogonally polarized light is proportional to the phase of Young's fringes. Thus, the retardation is equal to the phase change of Young's fringes before and after the insertion of the retarder into the common-path interferometer. The phase of Young's fringes is calculated from Fourier cosine and sine integrals of the fringe profile. The experimental results for wave plates, a Soleil-Babinet compensator and a Pockels cell are presented with error estimations. The accuracy of the retardation measurement is estimated experimentally to be higher than $\lambda/2100$.

1. INTRODUCTION

Polarization devices such as wave plates and compensators have been used in several optical systems such as interferometers, optical pick-ups, and light modulators.[1] Wave plates and compensators are important devices in polarization interferometers.[1] These devices induce a phase change between orthogonally polarized light in the directions of neutral axes of the devices, and this phase change is called as a retardation of light. A field-induced birefringence in a photorefractive or a photoconductive material is also significant which is used in the fileds of wave mixing, phase conjugation, and real-time signal processing applications.[2]-[4] The retardation of a polarization device can be measured by several methods such as Senarmont's[5] and Bruhat's methods[6] with a half-shadow plate or a quarter-shadow plate, heterodyne polarizing interferometry[7] and ellipsometry.[8]-[10] In Senarmont's and Bruhat's methods, wave plates and polarizers are mechanically rotated to match the light intensities to be equal, and the rotation angle of the polarizer gives the retardation of the material under study.[5],[6] In ellipsometers, it is also necessary to rotate a quarter-wave plate or a polarizer to calculate a retardation,[9],[10] even if electric phase modulation devices such as a Pockels cell are used.[8] In a heterodyne polarizing interferometer, an electric phase-meter gives the retardation directly. However, one or two acoustooptic light modulators or a Zeeman laser are needed to obtain a frequency-shifted light beam.[7]

In this paper, a new method for a highly precise measurement of a retardation is presented, in which a method for detecting a phase of Young's fringes is utilized in a common-path polarization interferometer.[11],[12] The experimental results of retardation measurements of wave plates, a Soleil-Babinet compensator(SBC), and a Pockels cell are presented with error estimations.

2. PRINCIPLE OF THE RETARDATION MEASUREMENT

A schematic diagram of a common-path polarization interferometer for a retardation measurement is shown in Fig.1. A linearly polarized light beam from a laser is magnified by an objective lens and impinges onto a birefringent wedge. The ordinary and extraordinary rays are deviated from each other after passing through the wedge, and equidistant and straight fringes, called Young's fringes in this paper, are formed on a linear image sensor.[12] The signals from the image sensor are processed by a computer before and after a retarder to be measured is inserted into the path of the laser beam. The polarization direction of the light beam impinging onto the retarder is rotated at 45° with respect to the neutral axes of the retarder. The principal axis (z-z´) of the birefringent wedge is parallel to one of the neutral axes of the retarder. The direction of the polarizer is rotated at 45° with respect to the principle axis of the wedge.

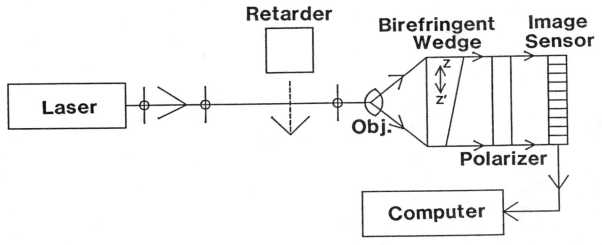

Fig.1 A schematic diagram of a common-path interferometer for a retardation measurement.

A profile of Young's fringes before the insertion of the retarder can be written as

$$I_1(\boldsymbol{x}) = W(\boldsymbol{x})\{1 + \cos(2\pi f \boldsymbol{x} - \varphi_1)\}, \tag{1}$$

where f and φ_1 are the spatial frequency and intrinsic phase of the Young's fringes. In this paper, the number of fringes in the detector area is called as the frequency of fringes. The window function of the fringe pattern which truncates the sinusoidal signal is represented by $W(\boldsymbol{x})$. The fringe profile after the insertion of the retarder is also expressed as

$$I_2(\boldsymbol{x}) = W(\boldsymbol{x})\{1 + \cos(2\pi f \boldsymbol{x} - \varphi_1 - \delta\varphi)\}, \tag{2}$$

where the retardation is represented by $\delta\varphi$. To detect the phases of the Young's fringes, Fourier cosine and sine integrals are calculated as

$$C_i = \int_{-1/2}^{1/2} I_i(\boldsymbol{x})\cos(2\pi f \boldsymbol{x})d\boldsymbol{x}, \tag{3}$$

$$S_i = \int_{-1/2}^{1/2} I_i(x)\sin(2\pi f x)dx, \qquad (i=1,2) \qquad\qquad (4)$$

where the range of the Young's fringes is normalized. The arctangent calculation with the integrals of C_i and S_i gives an estimated phase φ_i' as

$$\varphi_i' = \tan^{-1}(S_i/C_i), \qquad (i=1,2) \qquad\qquad (5)$$

where φ_2 stands for $\varphi_1 + \delta\varphi$. Therefore, the retardation can be obtained as the difference between two phases of φ_2 and φ_1. However, the phase error $|\varphi_i' - \varphi_i|$ is caused by the spatial truncation of the fringes. Other phase errors are caused by the determination error of the frequency f, and a random noise of the image sensor. These phase errors have been already analyzed using a Fourier transform of Eq.(1).[11] Assuming that (1): the window function is a Hanning, (2): the frequency of Young's fringes is greater than 10, (3): the frequency error is 0.1, and (4): the fringe contrast is 0.4, the phase error due to the data truncation is estimated to be less than $\lambda/3600$, where λ is the wavelength of light. When the S/N ratio of the image sensor is about 100, the total error is estimated about $\lambda/1000$. Thus, a retardation can be measured with a very high accuracy using the phase detection of Young's fringes.

3. EXPERIMENTS

A schematic diagram of an experimental setup is already shown in Fig.1. A linearly polarized light from a He-Ne laser(3.3 mW) is magnified by an objective lens($\times 50$), and passes through the birefringent wedge which is made of quartz and whose wedge angle is 12°. A fringe profile is detected by the image sensor (Reticon RL256G) which has 256 photodiode arrays. The maximum voltage of 5 V from the image sensor is quantized to yield 2048 levels (12 bit) using an A/D converter.

3.1 Stability and Accuracy of the Phase Detection

An output signal from the image sensor is shown in Fig.2(a), where the amplitude of the sinusoidal signal, which is denoted by α, is about 0.9 V and the fringe contrast is about 0.74. The frequency of Young's fringes is estimated as 10.0 using a digital Fourier transform. The intensity noise of the signal in Fig.2(a) is due to the internal reflection in the face plate of the image sensor and irregularities of sensitivities of the photodiodes. The difference between two successively digitized signals from the image sensor is shown in Fig.2(b), where the abscissa represents the pixel number of photodiode arrays and the ordinate represents the quantized levels. The peak-to-peak difference is about 4 levels and the standard deviation σ is about 1.02 levels, which corresponds to 2.5 mV. Thus, the signal to noise ratio S_N of the fringe, which is given as $\alpha/\sqrt{2}\sigma$, is equal to 255.[11] Therefore, the standard deviation σ_φ of the phase error due to the random noise, which is given by $1.22/\sqrt{N-1}S_N$, is equal to 0.017° and the minimum detectable phase change $3\sigma_\varphi$ becomes to be 0.052° ($\sim\lambda/7000$). The phase change measured for 50 sec is shown in Fig.3(a), where a Hanning window is used and the standard deviation is 0.026°. Thus, the minimum detectable phase change is 0.077° ($\sim\lambda/4700$). The measured minimum detectable phase change is about 1.5 times greater than that estimated theoretically. The phase error at low frequencies in Fig.3(a) is mainly due to a lateral position change between the birefringent wedge and the image sensor, whose distance is 200 mm. To evaluate the phase change at high frequencies in Fig.3(a), the difference between the phase averaged over 5 sample points and the original phase is calculated to eliminate the low frequency components, and the result is shown in Fig.3(b). The standard deviation is 0.015° and then the minimum detectable phase change is 0.045° (~λ/8000). This sensitivity obtained experimentally well coincides with that estimated theoretically.

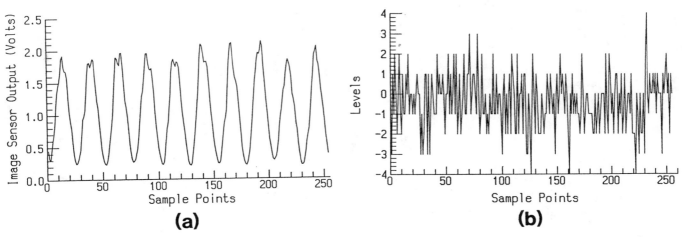

(a)   **(b)**

Fig.2 (a) Output voltage from the image sensor, and (b) the difference between two successively digitized voltages from the image sensor.

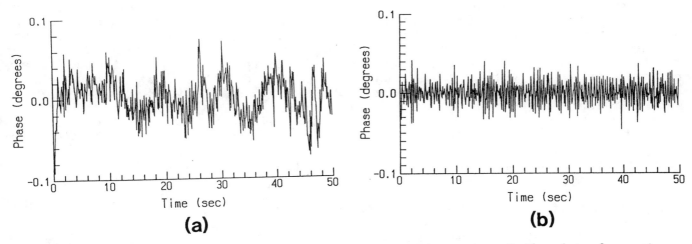

(a)   **(b)**

Fig.3 (a) Phase fluctuation for 50 sec in the stable state of the interferometer shown in Fig.1, where the standard deviation is 0.026°. (b) Phase difference between the phase averaged over 5 sample points and the original phase (a).

The phase change shown in Fig.3 is due to the random noise of the image sensor. The other phase error in the whole range of 2π is caused by the data truncation in the calculation of the phase, which can be estimated theoretically using a Fourier transform.[11] Assuming that the frequency of fringes is an integer number of 10, the frequency error is 0.1, and the fringe contrast is 0.74, the result of the error estimation is shown in Fig.4, where the abscissa represents the true phase φ and the ordinate represents the phase error $|\varphi' - \varphi|$. The maximum error shown in Fig.4 is 0.015° (~λ/24000), which is almost equal to the standard deviation of the phase error due to the random noise which is estimated theoretically or experimentally. Therefore, the total accuracy of the phase measurement is the sum of two phase

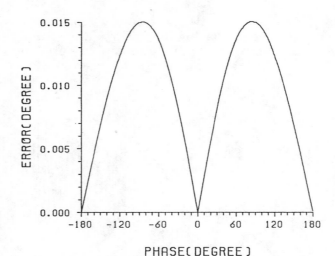

Fig.4 Calculation error due to a data truncation in the case that the frequency of Young's fringes is 10.0, the frequency deviation is 0.1, and the fringe contrast is 0.74. The abscissa represents the true phase and the maximum error is 0.015°.

errors, $6\sigma + |\varphi' - \varphi|$, and it becomes about 0.17° ($\sim \lambda/2100$).

The lowest limit of the sensitivity in the present method, $3\sigma + |\varphi' - \varphi|$, is estimated at 0.09° ($\sim 2.5 \times 10^{-4} \cdot 2\pi$), this sensitivity is comparable to that using the heterodyne detection technique with a Zeeman laser, which is $1.67 \times 10^{-4} \cdot 2\pi$.[7] The sensitivity in the ellipsometer using ADP crystals[8] reaches $4.6 \times 10^{-5} \cdot 2\pi$, which is 10 times greater than that in the present method. In Senarmont's method, the sensitivity reaches $5 \times 10^{-3} \cdot 2\pi$ without a half-shadow plate(HSP) and $5 \times 10^{-5} \cdot 2\pi$ with the HSP.[5] When a De Forest Palmer's compensator is used, the sensitivity reaches the minimum sensitivity among the methods using compensators, which is $5 \times 10^{-6} \cdot 2\pi$.[5] Thus, the present method has not so high sensitivity compared with another method, but the other birefringent devices such as wave plates, compensators, and field-induced materials are not needed except for the birefringent wedge. Therefore, a various wavelength can be used and the measuring system is very simple.

3.2 Calibration of a Soleil-Babinet Compensator

A Soleil-Babinet plate has been used well as a variable compensator in a common-path polarization interferometer, because the retardation of the SBC can be linearly changed by the moving distance between two birefringent wedges. To obtain the numerical value of the retardation, it is necessary to calibrate the retardation of the SBC with respect to the moving distance. The calibration can be performed easily using the present interferometer. The SBC is made of quartzes and the wedge angle is about 2.˚ One wedge of the SBC is successively moved at intervals of 1 mm, and data are acquired 16 times. The result is shown in Fig.5(a), where the abscissa is the moving distance and the ordinate is the retardation. This figure shows that the retardation is linearly proportional to the moving distance. The inclination factor of the line is obtained as 19.66°/mm using a least square line fitting. A very small difference in retardation can be induced by a precise control of the SBC movement.

To demonstrate the minimum detectable phase change, the SBC is moved 3 times at intervals of 10μm, in which 32 data are acquired every static state. The result is shown in Fig.5(b) where the abscissa represents the order number of the data and the ordinate represents the measured retardation. The phase change due to the 10μm movement of the SBC is estimated at 0.2° from the Fig.5(a), and we find clearly that the phase jumps occur at intervals of ~0.2° as shown in Fig.5(b). The maximum amplitude of the phase fluctuation in every static state in Fig.5(b) is ~0.1° which is almost equal to the estimated phase accuracy of 0.09°. Therefore, it is concluded that the total phase accuracy of $\lambda/2100$ can be reached sufficiently in the present technique.

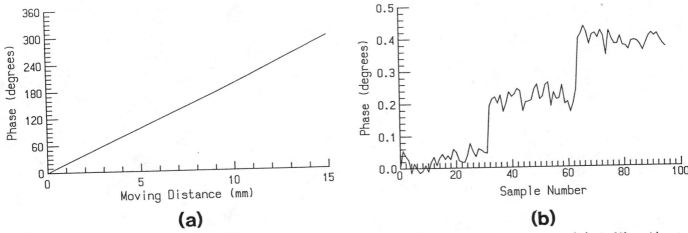

(a)　　　　　　　　　　　(b)

Fig.5 Experimental results for a Soleil-Babinet compensator(SBC).　(a) Calibration curve of the retardation of the SBC which is moved at the interval of 1 mm. (b) Retardation change for the successive movement of 10μm.

3.3 Retardations of Wave Plates

Half- and quarter-wave plates are very important devices in polarization systems such as interferometers and optical discs. The retardations of wave plates can be measured easily using the present technique.

The output voltages from the image sensor are shown in Fig.6(a), where lines a and b correspond to the data before and after the insertion of a half-wave plate(HWP) into the path of the interferometer, respectively. The phase difference between two phases of lines a and b is 176.83°, which corresponds to $\lambda/2.04$. When the HWP is rotated at 180°, the retardation is 181.06° which is equal to $\lambda/1.988$. The difference between two retardations is due to the position change of the measured points in the wave plate. A slight change of the thickness of a wave plate causes a small retardation change. However, we conclude that the measured wave plate is very close to a half-wave plate.

In the case of a quarter wave plate (QWP), the output voltages from the image sensor are shown in Fig.6(b), where data before the insertion of the QWP are plotted by line a. Signals after the insertion are plotted by lines b and c, where the QWP is rotated at 90° on its axis for line c. The retardations are measured as 254.1° and −96.6° for lines b and c, respectively. These retardations are corresponding to $\lambda/3.40$ and $-\lambda/3.73$, respectively. The difference between two retardations is due to the position change in the QWP by its rotation. This result shows that the wave plate measured is not especially a good quarter-wave plate.

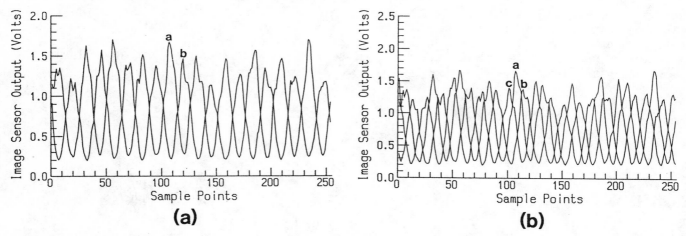

Fig.6 Output voltages from the image sensor, (a) for a half-wave plate, (b) for a quater-wave plate.

3.4 Retardation of Pockels Cell

A field-induced retardation measurement is significant in many application fields such as a light modulation, signal processings, and field sensings. A Pockels cell (Lasermetrics 3031FW) is measured as a representative field-induced retarder, which is consist of four ADP crystals in order to cancel out the intrinsic retardations of the crystals. All of the ADP has the same dimensions of 26 mm in length along the optical path and 3 mm in height between the electrodes. To obtain a half-wave voltage, the retardation is measured as a high voltage is applied to the Pockels cell(PC). The result is shown in Fig.7(a) where the retardations for the increase and decrease of the applied voltage are plotted by lines a and b, respectively. The half-wave voltage is obtained as 296 V from this figure. A hysteresis of the retardation is very small in the range of a higher voltage, but a hysteresis can be observed in the range lower than 40 V. To clarify the hysteresis, a small retardation is measured as a lower volatge of ±10 V is applied to the PC. The result is shown in Fig.7(b), where the order and the increase and decrease of the voltages applied to the PC are shown by the numbers and arrows near the curves, respectively. The hysteresis of the retardation is clearly observed in this figure, and the results show that the field-induced retardation can be measured continuously with a high accuracy.

4. CONCLUSION

An application of a phase detection of Young's fringes and a common-path polarization interferometer are presented for the case of a retardation measurement. From the accuracy evaluation of the phase measurement in the present method, the minimum detectable phase change is $\sim\lambda/4000$ and the accuracy is $\sim\lambda/2100$. To demonstrate the accuracy of the phase measurement, retardations of several representative objects of a Soleil-Babinet compensator, wave plates, and a Pockels cell are measured, and the results show that the retardation can be obtained with a high accuracy more than $\lambda/2100$.

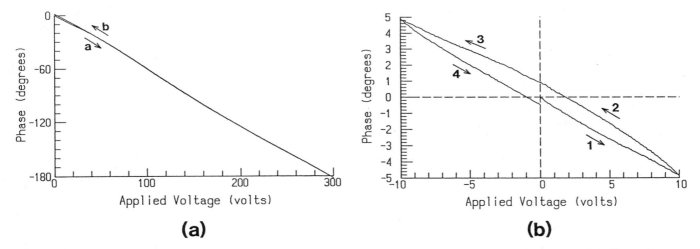

Fig.7 Experimental results for a Pockels cell(PC).   (a) Retardation as a high voltage applied to the PC in order to obtain the half-wave voltage.   (b) Hysteresis of the retardation in the range of ±10 V.

The present interferometer with the birefringent wedge is a very simple common-path interferometer, and a variable wavelength of light can be used in the same configuration of the interferometer.

The author would like to thank I. Yamaguchi of Institute of Physical and Chemical Research for his helpful discussions.

5. REFERENCES

1). M. Francon and S. Mallick, *Polarization Interferometers: Application in Microscopy and Macroscopy*, Wiley, London, 1971.
2). I. P. Kaminow and E. H. Furner, "Electrooptic Light Modulators," Proc. IEEE **54**, 1374(1966).
3). A. J. Fox and T. M. Bruton, "Electro-optic Effects in the Optically Active Compounds $Bi_{12}TiO_{20}$ and $Bi_{40}Ga_2O_{63}$," Appl. Phys. Lett. **27**, 360(1975).
4). J. P. Huignard, H. Rajbenbach, Ph. Refregier, and L. Solyman, "Wave Mixing in Photorefractive Bismuth Silicon Oxide Crystals and its Applications," Opt. Eng. **24**, 586(1985).
5). H. G. Jerrard, "Optical Compensators for Measurement of Elliptical Polarization," J. Opt. Soc. Am. **38**, 35(1948).
6). G. Bruhat, *Optique* Masson, 1959.
7). L. Yao, Z. Zhiyao, and W. Runwen, "Optical Heterodyne Measurement of the Phase Retardation of a Quarter-Wave Plate," Opt. Lett. **13**, 553(1988).
8). H. Takasaki, "Photoelectric Measurement of Polarized Light by Means of an ADP polarization Modulator, I. Photoelectric Polarimeter," J. Opt. Soc. Am. **51**, 462(1961).
9). R. M. A. Azzam and N. M. Bashara, *Ellipsometry and Polarized Light*, North-Holland, Amsterdam 1980.
10). P. S. Hauge and F. H. Dill, "Design and Operation of ETA, an Automated Ellipsometer," IBM J. Res. Dev. **17**, 472(1973).

11). S. Nakadate, "Phase Detection of Equidistant Fringes for Highly Sensitive Optical Sensing, I. Principle and Error Analyses," J. Opt. Soc. Am. **A5**, 1258(1988).

12). S. Nakadate, "Phase Detection of Equidistant Fringes for Highly Sensitive Optical Sensing, II. Experiments," J. Opt. Soc. Am. **A5**, 1265(1988).

Hexflash phase analysis examples

Lawrence Mertz

Lockheed Research Laboratory, Dept.97-20
3251 Hanover St., Palo Alto, CA 94304

ABSTRACT

A hexflash phase analyzer is an electronic circuit that solves the common phase from a three-phase input. Two applications are described, one being an interferometric angle encoder having unprecedented resolution, and the other being a real-time television fringe pattern analyzer.

1. HEXFLASH CIRCUITRY

A hexflash circuit uses six flash analog-to-digital converters connected to the six permutations of a three-phase signal (three buckets) to resolve the basal phase in digital format. The input signal expressions are

$$A = \beta + \gamma \cos(\phi - 120°) ,$$
$$B = \beta + \gamma \cos(\phi) ,$$
$$C = \beta + \gamma \cos(\phi + 120°) ,$$

and the circuit solves for ϕ, independent of β and γ. It is described in reference 1, and is like a synchro-to-digital converter except that it is not limited in speed by any a.c. carrier excitation. Figure 1 diagrams the circuit. The circuits built so far deliver a digital byte for a resolution of 1/256 cycle with submicrosecond sampling.

The output of the circuit conveniently adapts to a digital phase-unwrapping and averaging filter (called a polar innovations filter), employing an adder and a multiplier-accumulator as shown in Figure 2, to gain even higher resolution over an unrestricted number of cycles.

2. INTERFEROMETRIC ANGLE ENCODER

When all the circuitry was introduced,[1] the input signals were obtained from a lenticular optical homodyne receiver, which slices a pattern of parallel fringes into thirds of a fringe to obtain the three phase-shifted signals all at once. Originally, the fringe pattern was generated by an interferometer having deliberately low sensitivity. Replacing that interferometer with one of high sensitivity as shown in Figure 3 leads to a superb angle encoder. This is really a Michelson interferometer in which the beams are folded by a pair of parallel mirrors that are mounted on a module that serves as the mobile rotor. The nifty feature is that the device remains in interferometric quality alignment for severe perturbations of all six degrees of freedom of the rotor, and that the path difference is insensitive to five of those degrees while being very sensitive to one angle of the rotor. One million fringes per radian result from a 12-cm spacing of the parallel mirrors and HeNe laser wavelength.

The interferometer adapts to the lenticular receiver by realigning the beamsplitter so that the return beams become separated and subsequently recombined by a lens to provide fringes on the lenticular receiver. The rotor module of the actual interferometer has a range of 18°, is counterbalanced, and dangles from a nylon fishline located at the symbol ⊗ in Figure 3.

The dynamic performance of the entire encoder is illustrated in Figure 4 where selected output bits are displayed from a digital-to-analog converter. Part A shows 16 fringes of phase, and that range could be easily extended with a counter. Part B shows phase for individual fringes, where the modest and reproducible waviness is the result of somewhat mismatched signals from the lenticular receiver. For a more detailed look, part C shows just the three least significant phase bits. Discrete microsecond time steps and the discrete 4-nanoradian phase steps are both conspicuous. The nonuniformity of the phase steps is inherent in translating the code to the best straight line, but notice that it is reproducible from fringe to fringe on the successive traces in C. Part D is like part C but includes averaging with an exponential decay constant of four samples in the digital filter so that the phase steps are reduced to one nanoradian. To avoid confusion, all parts of Figure 4 are shown with the rotor moving in the same direction, but the encoder operates just as well in the other direction with the phase decreasing as a function of time.

3. TELEVISION ANALYSIS

For interferogram analysis, three-phase signals may be obtained from pixels of a television format. In this case the interferogram is adjusted to give nominally vertical fringes with three pixels per fringe. The pixel intensities are stored sequentially in three sample-and-hold amplifiers for transfer to the hexflash circuit. Each new pixel replaces one of the three samples in cyclic order, and the phase is updated with each new pixel. A video digital-to-analog converter offers real-time video display of the phase. There is less than a microsecond delay in the analysis, so the phase signal is useful for dynamic adaptive optics control applications.

Figure 5 shows snapshots of a TV monitor in which the left images were inputs for the corresponding right images. Although these are just simulated interference fringes, the circuitry doesn't care. The uppermost pair of images includes a circular region where the fringes have reverse contrast to illustrate the capabilities of the phase analysis through boundaries of low contrast. The middle pair illustrates a more typical pattern.

The lowest pair illustrates the operation for a line that the analyzer interprets as an individual fringe. The phase specifies the abscissa of the peak for the cosine bell curve fitting through the extremum pixel and its two neighbors. In contrast to a center of gravity, the phase-defined centroid is virtually unbiased. For the actual circuit, the phase resolution is 1/256 over three pixels, or 1/85 pixel. This centroiding application could be used to specify the positions of spectral line features, grid lines, fiducial marks, Hartmann spots, or flare events.

<u>REFERENCE</u>

1. L. Mertz, "Optical homodyne phase metrology," Appl. Opt. **28**, 1011 (1989)

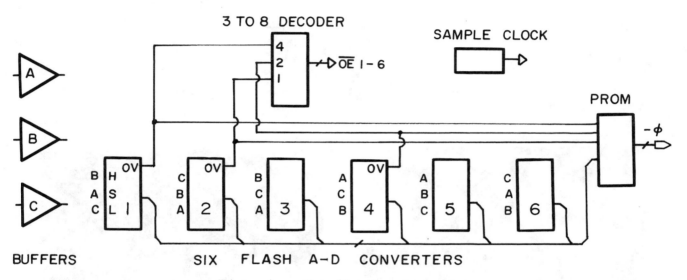

Fig. 1.   Hexflash circuit.

$$\Phi_{t+1} = \Phi_t - N^{-1}(\Phi_t - \phi_t)_R$$

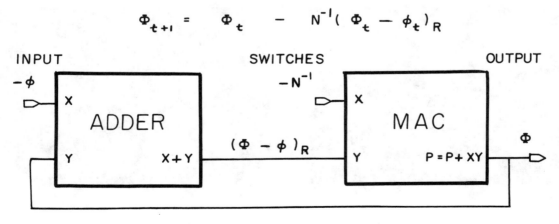

Fig. 2.   Polar innovations filter.

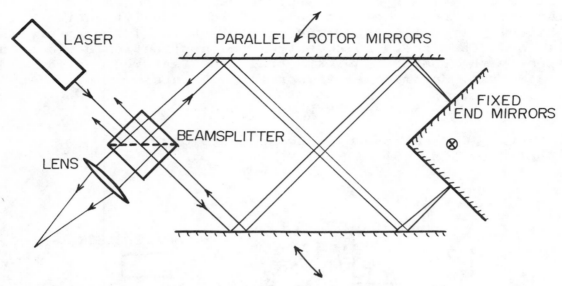

Fig. 3.   Angle-sensitive interferometer.

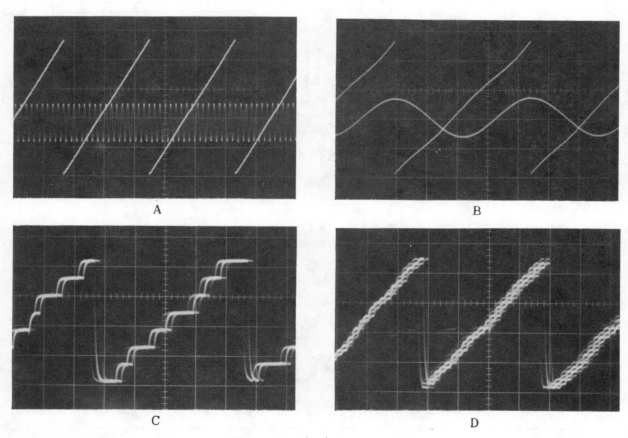

Fig. 4.   Angle vs. time.   (see text)
Horizontal   A, 10ms/div;   B,   1ms/div;
C,   5μs/div;   D,   5μs/div.

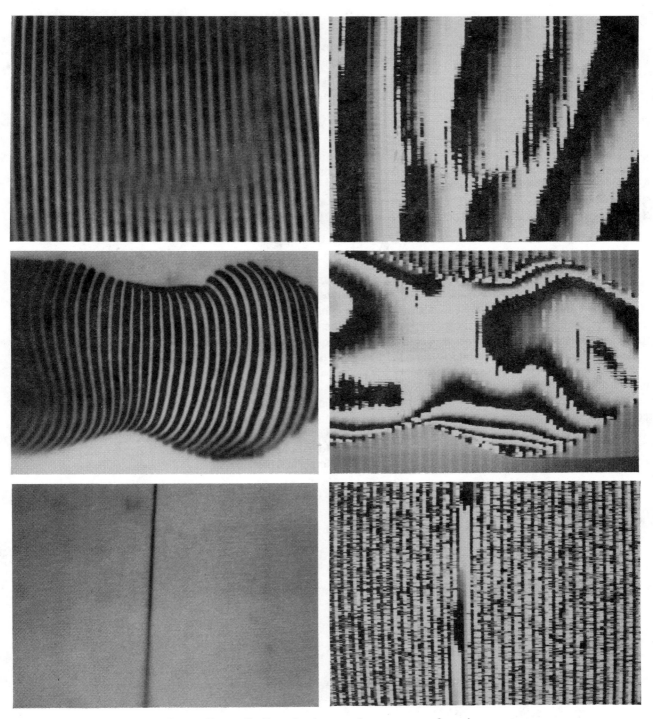

Fig. 5. Television phase analysis.

High speed, large format wavefront sensor
employing hexflash phase analysis
Jimmy Roehrig, Paul Ehrensberger, Mark Okamura
Lockheed Research Laboratory
3251 Hanover St., Dept 91-10
Palo Alto, Ca 94302   (415)424-4022

ABSTRACT

The concept of hexflash phase analysis described in a previous talk[1] has been extended to two dimensions. The associated phase unwrapper also works in two dimensions, making the device useful for high speed wave front sensors. A wave front sensor is described which uses high speed 32 x 32 arrays at each focal plane, enabling real time phase measurment at up to 1024 points. These measurements are acquired at a nominal frame rate of 5 KHz. The processing electronics which will be described adds a transport delay of less than 1/20th of a frame.

1.WAVE FRONT SENSING

Wave front sensors in use today are usually variations of two fundamental catagories: Hartmann sensors and interferometric measurements. Hartmann sensors, which measure an array of x and y slopes by spot centroiding, are very simple conceptually, and are capable of very high dynamic range measurements. Their disadvantages are that centroiding a large array of spots usually requires very heavy processing capabilities, and/or involves very exotic optical components for optical processing. Alignment of these devices is also very difficult, and can result in a loss of dynamic range. In contrast, the interferometric technique described in this paper involves only relatively common optical components, is easy to align, involves very simple electronic processing, and is easily expandable to any number of channels by selecting the appropriate focal plane. A disadvantage that all interferometric techniques share is the need to perform phase unwrapping.Performing phase unwrapping in 2 dimensions in real time imposes a strong requirement for simplicity and ease of hardware implementation that eliminates many well known robust algorithms from consideration. We demonstrate a solution to this problem in this paper. Finally, the algorithm we will describe is very general - it can be employed with any interferometer (e.g. holographic shearing, Mach-Zender, etc). This paper describes one optical and electronic implementation that is simple and particularly useful.

2.ALGORITHM

An interferometer of any type produces a modulated intensity of the form:

$$I_1 = a_0 + a_1 \cos(\phi) \tag{1}$$

If a phase retardaton of $\lambda/3$, $2\lambda/3$ can be introduced into the reference beam the intensity would become:

$$I_2 = a_0 + a_1 \cos(\phi + \lambda/3) \tag{2}$$

$$I_3 = a_0 + a_1 \cos(\phi + 2\lambda/3) \tag{3}$$

respectively. A small amount of trigonometric manipulation with eqs. 1-3 yields the result:

$$\phi(x,y) = \mathrm{atan}\{(I_0 - I_2)/(I_1 - I_2)\} \tag{4}$$

within a constant phase. Other values of phase retardation achieve the same results. $\pi/4$, $\pi/2$ are most often used. Bareket[2] described a device using these values of phase retarders. A circuit that calculates the ratio in equation 4 and performs the arctangent will yield phase. There are many devices that use this technique. An example is the Plessy chip ("Pythagorean processor") that calculates the arctangent. Since difficulties with this arctangent calculation are well known, we will now describe how our algorithm exploits the features of interferograms with 120 degree relative retardation.

Fig. 1 shows three sinusoids separated by 120 degrees, representing the three interferograms eq. 1-3. The phase is plotted on the abscissa, the intensity on the ordinate. Note that for any phase value, the difference between the highest and lowest intensity is approximately constant, while the intermediate intensity goes from the lowest to the highest in a 60 degree interval. Inspecting the interval from 0 to 60 degrees, where a,b,c represent the highest, intermediate, and lowest intensities respectively, clearly a one to one relationship exists between (b-c)/(a-c) and the value of the abscissa, ϕ. By using (b-c)/(a-c) to address a prom containing the corressponding values of ϕ, a measurement of the 3 interferogram intensities at each point of the focal plane is converted directly to ϕ modulo 60 degrees.

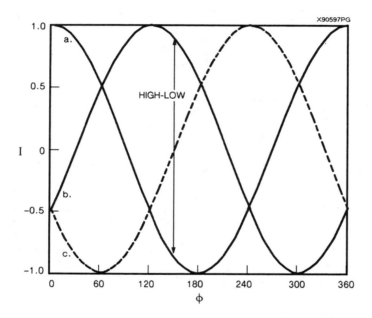

Fig.1 Intensity distributions for retardation
of 0,120, and 240 degrees.

Since (int-low)/(high-low) gives the phase modulo sixty degrees, it remains to find which 60 degree interval the phase belongs to. This is given by determining the correct permutation of a,b,c that gives high, int, low intensities respectively. Note in fig.1 that each 60 degree interval from 0 to 2π coresponds to a unique permutation of a,b,c. Thus the phase from to 0 to 2π is determined. The only remaining ambiguity occurs when the phase increases beyond 360 degrees, and drops back down to 0. By requiring that the phase difference between adjacent pixels is less than $\lambda/2$, a 'turns counter' that counts the number of such occurances and makes the appropriate correction will complete the determination of the phase. This is exactly the same technique as used by Mertz[1] in his single line unwrapper. A 2-dimensional interferogram requires slightly more sophistication and again we exploit a feature of the 120 degree separated interferograms. It is necessary to know what region of the focal plane contains valid data, and which regions do not (e.g. the dark corners when a circular interferogram is imaged onto a square array). With relative retardation of $\lambda/3$, the difference hi-low is approximately constant and equal to a characteristic value. Alternatively, the sum of all 3 intensities eqs. 1-3 is just the sum of the single beam irradiance, approximately constant. A each row of 3 focal planes is read out, one of the quantities is compared to a threshold level to determine the first pixel containing valid data. The value of the phase at this pixel is stored in a holding register. In the next row, the first valid pixel is required to be within $\lambda/3$ of this value. In this way the 1 dimensional phase unwrapping spreads out into 2 dimensions. This works well whether the edge connecting first valid pixels is straight or curved like the circumference of a circle, as long as the edge is continuous.

3. OPTICS

Fig. 2 shows a diagram of the optics. BS1 first splits the incoming beam into two arms to measure the x and y sheared phase. The main component in each arm is a shearing interferometer consisting of a polarizing beam splitter cube and 2 mirrors arranged in a triangle. The incoming beam is polarized at 45 degrees so the intensity in the two arms of the interferometer are equal. A shear of the two beams can be introduced either by a rotation (in the plane of the triangle) of either mirror, or the cube beam splitter. For alignment reasons the cube bs angle is used to introduce shear of the p beam with respect to the s beam. Rotation of the mirrors or cube out of the plane of the triangle will introduce not only a shear in the perpendicular direction, but a relative tilt in the opposite plane, and are thus avoided. In order to introduce y shear as well as x, an identical system in mounted in the perpendicular plane.

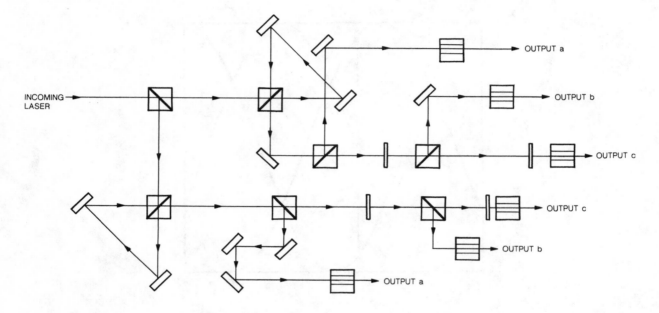

Fig. 2 Wave Front Sensor Optics

The sheared beams are then sent through two nonpolarizing beam splitters to produce 3 output paths. One path has no retardation plates and thus produces the main interferogram-camera a at the analyzer, the next path passes through one $\lambda/3$ retarder producing camera b at the analyzer, the final one passes thru two $\lambda/3$ retarders to produce camera c. Each view thus has three focal planes that produce the interferograms described in equations 1-3 above.

4.ELECTRONICS

The electronics is designed to implement the algorithm described in section 2 with as little transport delay as possible. At a frame rate of 5 Khz, transport delay of less than 1/20th of a frame was achieved.

The processing system is partitioned into six subsections: cameras; single-channel processors (SC); phase un-wrapper(PU); interfaces; global timing and control; and microvax 3200 host computer (see Fig. 3).

4.1 Camera

To minimize cost and complexity, the system uses raw camera data without individual pixel correction. Nevertheless, with pixel-to-pixel gain and offset variations less than 10% of full scale, reasonable phase resolution can be achieved.

The cameras are composed of an RA6464N imaging array (64x64 pixels), analog signal conditioning circuits, local timing and control, and high-current buffers.

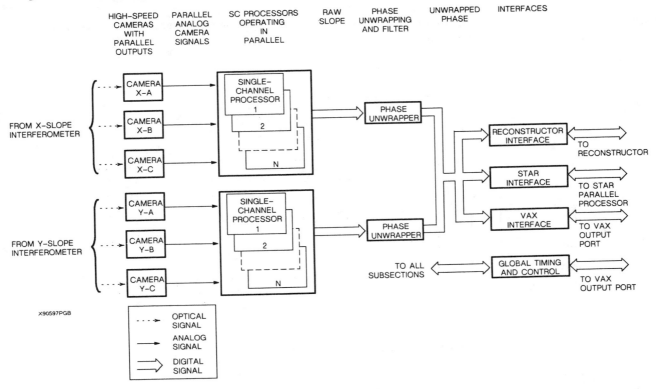

Fig.3 Block digram of wave-front sensor electronics.

The Reticon RA6464N imaging array was chosen because its parallel outputs make high frame rates possible. It has 32 parallel outputs connected so that 32 32x4 pixel subarrays are read out simultaneously. This topology is well matched to our requirements since it permits using three of the four quadrants (8 each of the 4x32 subarrays) as independent focal planes, and thus allows the use of only two cameras (one each for x and y phase)in a 1024 subaperture system. These three focal planes in each view are combined into one camera with a 5 ft fiber optic bundle with the input end split into separate quadrants, and the single output end mated to the RA6464N. Analog signal conditioning is performed by a simple passive integrator with reset. An analog correction scheme is used to reduce gain and offset errors to required values within each 4x32 subarray.

Local timing and control provides the multiple timing signals needed to operate the focal plane and signal conditioning circuits. The cameras were mounted separately from the other electronics and high current buffers allow up to 100 feet of cable between them.

4.2 Single-channel processors (SC).

The camera outputs are routed to the SC. The flash ADCs on the SC (see Fig.4) calculate all 6 permutations of (b-c)/(a-c) to yield the phase modulo 60 degrees. This configuration of 6 flash ADCs has been given the name of Hex-flash conversion. Which permutation of a,b,c is correct can be derived from the 3 overflow bits which give the correct order of a,b,c. These overflow bit are sent to a decoder which output enables the correct flash ADC. The 3 bits also act as a grey-code address for the upper address for the phase (ϕ) PROM. The raw slopes from the PROM are sent to the PU section.

Another function is available. A mode may be selected which allows one raw image plane to be digitized and sent to the computer for calibration or diagnostic purposes.

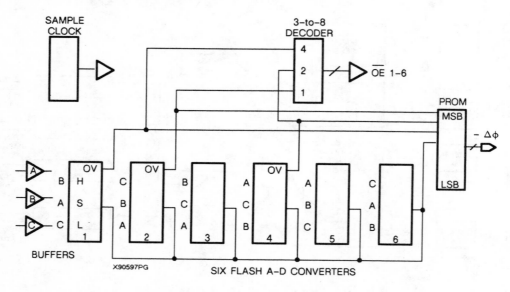

Fig.4 Single channel processor hexflash ciruit.

4.3 Phase-unwrapper (PU).

At the input of the PU board (Fig.5) the data is reformated to a single stream such that it is arranged in full column order. The phase unwrapping implementation is relatively simple, but the operation is complex.

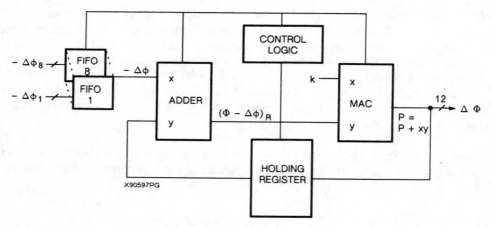

Fig. 5 Phase Unwrapper.

Automatic phase unwrapping in one dimension is intuitively simple: detect the discontinuity due to the inherent modulo 2π behavior and simply add or subtract 2π to obtain the unwrapped phase. Our implementation makes use of the difference between the present and last sampled phase to detect and correct (unwrap) the phase. This algorithm is extended to two dimensions by requiring the first phase calculated in a new column to be within $+-\pi$ of the first phase of the previous column. This is reasonable because of the close proximity of these pixels. Fig. 5 shows the hardware implementation of this function. Assume that the constant k=-1., and that the delay latch is enabled. The dynamic range of 8 bits is equivalent to 2 pi. The 8 bit inputs to the adder, the negative of the local phase -φ and the least significant 8 bits (x) of the output Φ are added. The resulting 8 bit number (carry ignored) is then passed to the multiplier accumulator.The phase is then given by the equation:

$$\Phi n = (\phi n - \Phi n\text{-}1)\bmod 2\pi + \Phi n\text{-}1 \tag{5}$$

Two-dimensional unwrapping is accomplished as described in section 2. The hold register latches the first valid phase at the beginning of each new column and is then enabled when the first valid phase of the next column is encountered. The 12-bit output allows up to eight "turns" in either direction. 4.4 Interfaces and global-timing and control (GT&c) sections. The slope interface provides high-speed interfacing to the controler portion of control system. Firmware options for spacial decimation of data is provided to this board. The host interface is a slow speed port to the uVAX 3200 workstation used to acquire image data as a single frame snap shot of slope data. Finally, GT&C provides syncronization of all subsystems and interprets commands from uVAX to WFS.

5. RESULTS

Fig. 6 shows the 3 interferograms (one view) for a wavefront that contains several waves of pure focus. Fig. 7a shows the reconstructed phase ramp, Fig. 7b is a plot through the central column.

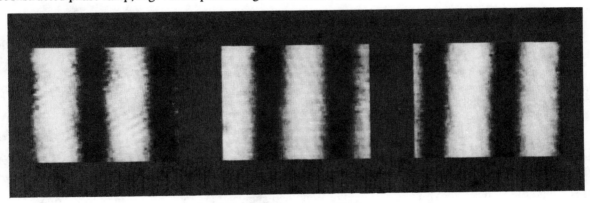

Fig. 6 Raw interferograms for focus.

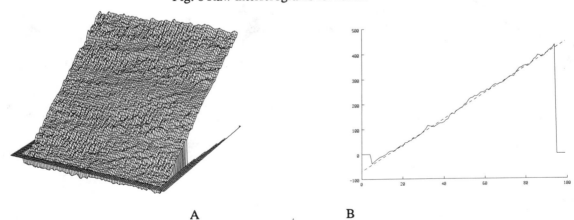

A   B
Fig.7 Reconstructed phase ramp.
A, surface plot;   B, crossection through middle column.

We observe in Figs. 6-7 that the phase ramp is consistant (see dotted line Fig. 7b) with that from pure focus. The observed rms deviation is $\lambda/38$. Some systematic effects are still visable, and are a function of alignment and calibration of the device, which is continuing to improve. An idea of the ultimate dynamic range and sensitivity of the device given ideal calibration can be obtained from the following considerations. Since phase unwrapping requires each pixel has a phase within $\lambda/2$ of the previous phase, this places a constraint on the maximum dynamic range, expressed as a maximum value of $\left| \delta\phi/\delta x \right|$ for a given shear Δx.

$$\Delta\phi = \delta\phi/\delta x \, \Delta x < \pi \tag{6}$$

For example, if it is required that a maximum slope of $\left| \delta\phi/\delta x \right| = 3$ waves/subaperture be measured, then if the system be required to measure 252 subapertures with the present array size of 32x32, we have :

$$\left| \delta\phi/\delta x \right| < 1.5 * 2\pi \tag{7}$$

then eq.6 implies that $\Delta x = 1/3$ pixel. We then ask what sensitiviy exists for this amount of shear. Eq. 6 implies:

$$\left| \delta\phi/\delta x \right| > \sigma/\Delta x = 3\sigma/\text{pixel} \tag{8}$$

With the algorithm described in section 2,

$$\sigma^2 \sim 2.67 * \sigma\text{int}^2 \tag{9}$$

where σint is the noise in each measured interferogram.

If we assume each interferogram has 3% noise (reasonable value for camera noise plus systematic effects), then $\sigma \sim$ .049 in units of $\lambda/6$, or $\sigma \sim .00817\lambda$. Hence:

$$\left| \delta\phi/\delta x \right| = 3(.00817)\lambda/\text{pixel} \sim \lambda/42 \tag{10}$$

6.ACKNOWLEDGMENTS

The authors would like to thank the computer support of Ron Sharbaugh and Dzung Nguyen, and Steve Triebes for his computer simulations.

7. REFERENCES

1. L. Mertz , "Hexflash phase analysis examples," this conference. See also L. Mertz, "Optical homodyne phase metrology,", Applied Optics, Vol.28, No.5, pg 1011, March 1989.

2. N. Bareket, "Three channel phase detector for pulsed wavefront sensing," Proc. SPIE 551, pg.12, 1985. See also N. Bareket, US Patent # 4,583,855 , 1986.

New image processing algorithms for the analysis of speckle interference patterns

H.A. Vrooman and A.A.M. Maas

Delft University of Technology, Faculty of Applied Physics
P.O. Box 5046, 2600 GA Delft, The Netherlands

ABSTRACT

A sequence of algorithms for processing interference patterns generated by a phase-shifting speckle interferometer is discussed. The processing yields the computation of in-plane displacement and strain components on the surface of an object, using a phase-shifting algorithm to compute the phase. Accurate phase measurement on a 512*512 grid is achieved by pixel-synchronous digitizing of four interference patterns and subsequent calculation of the phase modulo 2π radians, using a two-dimensional look-up table. A pipeline of Datacube image processing modules is configured to perform this measurement. Digital image processing algorithms have been developed for phase unwrapping, phase restoration and smoothing. During these processing steps, invalid pixels due to low modulation or saturation are neglected. For phase unwrapping an algorithm has been developed that, starting at a chosen start pixel, propagates a "wavefront" of unwrapped phase data through the data set and that takes a set of neighbouring pixels into account to detect 2π steps. Subsequently, phase restoration is done by averaging valid neighbouring pixels. Basic binary image processing techniques are used to solve the problem of irregularly shaped objects due to holes and shadows. The measured phase change is used to compute the in-plane displacement and strain components of the deformed object. Results of a measurement of in-plane displacement and strain components on the surface of a simple aluminium object are shown.

1. INTRODUCTION

Speckle interferometry is a non-contact measuring technique that can be used to determine the displacement at points on the surface of a diffusely reflecting object with an accuracy in the order of a fraction of the wavelength of the light used in the interferometer. This technique can be used for a variety of purposes, e.g. measurement of deformation, strain and shape of mechanical components as well as detection of defects, cracks and disbonds in materials. Although a number of applications only require a qualitative measuring technique, a quantitative technique like digital phase-shifting speckle interferometry (DiPSSI) using digital image processing techniques offers obvious advantages in the case of strain analysis. In 1985, the use of a phase shifting technique for direct phase evaluation of speckle intensity patterns[1,2] was reported.

Digitizing the interference patterns and using a reference beam phase-shifting technique allow the calculation of the phase modulo 2π at each point on the surface of the test object. Because of the random phase distribution in a speckle field, the calculated phase values have to be subtracted from the corresponding values of a reference phase measurement to give a deterministic result. Without further processing, this result is sufficient if a qualitative inspection of the object is required. However, in the case of strain analysis, where the calculation of the complete 3D displacement vector field as well as the in-plane strain components is required, further processing is necessary.

If the object is properly imaged onto the detector, the calculated phase change gives a measure of the displacement component in the direction of the sensitivity vector of each object surface point[3]. By performing at least three measurements, with different sensitivity vectors, the complete three-dimensional displacement vector can be calculated in principle. The in-plane strain components can be obtained by calculating the first derivatives of the in-plane displacement components[4,5]. The results presented in this paper refer to an object surface, lying in the two-dimensional space.

2. PHASE MEASURING METHOD

The intensity distribution $I_i(x,y)$ of the i-th interference pattern generated by a phase-shifting speckle interferometer can be expressed as:

$$I_i(x,y) = I_B(x,y) + I_M(x,y) \cos[\phi_s(x,y) + \phi_o(x,y) + i\Delta\phi], \qquad (1)$$

where I_B is the background intensity, I_M the modulation amplitude, ϕ_s the random speckle phase, ϕ_o the phase difference between the object and reference beams without speckle, $\Delta\phi$ the reference beam phase shift, (x,y) the position in the image (omitted in Eqs. 2 and 3) and i an integer number indicating the i-th interference pattern[1,2,6]. To calculate $\phi_s + \phi_o$, a method using four phase shifted interference patterns with $\Delta\phi = \pi/2$ radians is used. It can be shown that $\phi_s + \phi_o$ can be computed modulo π by:

$$\phi_s + \phi_o = \arctan [(I_4 - I_2) / (I_1 - I_3)]. \tag{2}$$

By considering the sign of the numerator and the denominator in Eq. 2, $\phi_s + \phi_o$ can be determined modulo 2π radians. Assuming that the speckle phase ϕ_s does not change[7,8], subtracting the result of a reference phase measurement yields the phase change ϕ of the object beam.

3. COMPUTER SYSTEM AND OPTICAL SET-UP

The experimental system (Fig. 1) consists of a speckle interferometer coupled to a VME/MC68000 microcomputer, used for digitizing and processing the interference patterns and for controlling a mirror position in the optical arrangement. The object illumination direction can be altered by alternatively blocking one of the illuminating beams. An Argon ion laser is used at an output power of approximately 100 mW and a wavelength of 514 nm. Phase-shifting is achieved using a mirror in the reference beam mounted on a Burleigh piezo-electric translator. The object used for studying the characteristics of the experimental system is a T-shaped aluminium plate (Fig. 2). A deformation can be introduced by putting a weight on the tip of the arm of the object.

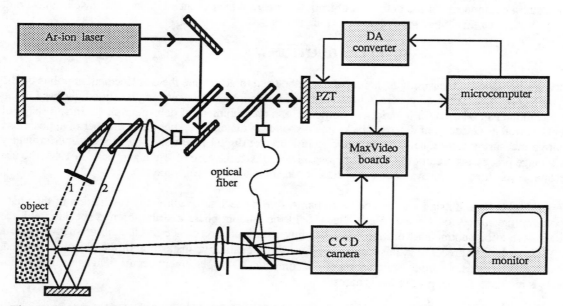

Fig. 1. Experimental system for 3D displacement measurement.

The microcomputer is equipped with a number of Datacube image processing modules. Fig. 3 shows the configured pipeline to measure the phase change ϕ modulo 2π radians. The pipeline consists of the following modules:

- one framegrabber to digitize the interference patterns (MaxScan SC).
- a 2.0 Mbyte and a 0.5 Mbyte ROI-stores (ROI = Region Of Interest) to store arrays of pixels (RS0.5 and RS2.0).
- two signal processors to perform simple arithmetic on pixel streams (MaxSp's MS1 and MS2).
- one multiplexer containing a 256*256 8 bit hardware look-up table (MaxMux MU).
- one display module for displaying the results and for further graphical needs (MaxGraph GH).

The initialisation of each module is done by the Ironics host computer. The CCD video camera, with 575*604 pixels, is pixel-synchronously coupled to the Datacube MaxScan framegrabber. This means that the pixel clock of the CCD camera and the sample clock of the framegrabber are equal (about 11.25 MHz). The elimination of video jitter thus achieved causes a significant reduction of the noise in the digitized interference pattern. A part of the image from the camera is digitized into a sub-array of 512*512 8 bit pixels and sent from one module to another through the MAXbus with a pixel rate of 10 MHz. Pixel stream delays due to each processing module are compensated with several delay lines on the modules. With this pipeline each 240 ms (6 video frame times) a new phase change modulo 2π radians is calculated. Fig. 4 shows the time-space scheme of the computation. The pixels of the used camera have an aspect-ratio of about 4:3 (H:V). The computed grid of phase changes ϕ is corrected by interpolation in horizontal direction to get the correct phase values.

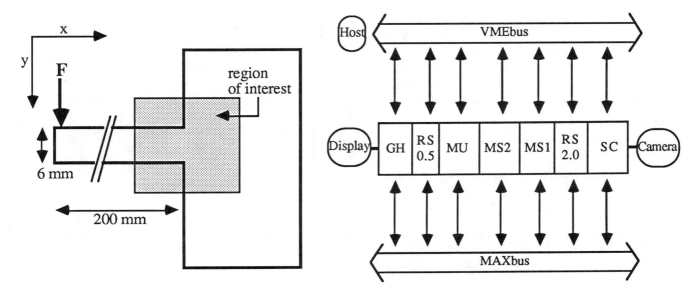

Fig. 2. T-shaped aluminium object.

Fig. 3. Datacube modules to measure the phase change on a 512*512 grid in 240 ms.

After the phase measurement on a 512*512 grid, further computations are done using general purpose computers. The image processing software environment is TCL-image, a software package containing basic image processing routines[15]. For time-consuming calculations, e.g. floating point operations, a SUN3/MC68020 workstation with a MC68881 co-processor was used. Paragraph 4 to 6 describe the processing steps after the measurement of the phase change ϕ.

4. DETECTION OF INVALID PIXELS

Evaluation of $\phi_S + \phi_0$ from Eq. 2 is performed by calculating the numerator and denominator values and then applying a 2D look-up table, that outputs the phase value modulo 2π. In the result image of this phase measurement invalid pixels, i.e. pixels having an inaccurate greyvalue, may occur. These pixels have to be identified, since they do not contain relevant data and will disturb the phase unwrapping and further processing of the relevant data. Characteristic for digital phase-shifting speckle interferometry is the loss of accuracy in points with low modulation caused by dark speckles and in saturated pixels caused by bright speckles. As can easily be derived from Eq. 1 the modulation amplitude I_M is:

$$I_M = \frac{\sqrt{(I_4 - I_2)^2 + (I_1 - I_3)^2}}{2}. \tag{3}$$

This corresponds to a circle with the centre at the origin of the $((I_4 - I_2),(I_1 - I_3))$ plane and a radius $2I_M$. So, pixels with a modulation amplitude below a certain threshold are detected using a circular area in the 2D look-up table with a radius representing the modulation amplitude threshold (Fig. 5).

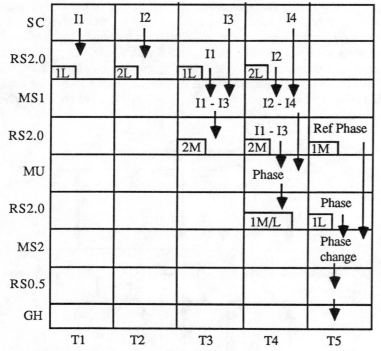

Fig. 4. Time-space scheme of the computation of the phase chage φ in 6 video frame times (T6 is used to stabilize the mirror). The code in the bottom left corners indicate the position of the data on the processing module.

Fig. 5. Application of a 2D look-up table to compute the arctangent.

Pixels having the maximum greyvalue 255 in at least one of the four phase-shifted interference patterns are assumed to be saturated and thus invalid. The detected invalid pixels are stored in a binary image B_m. This image is used as a mask during further processing steps. The image itself is also processed during phase unwrapping and phase restoration to identify object and background areas.

5. PHASE UNWRAPPING AND RESTORATION

Because the phase change φ is calculated using an arctangent, 2π-steps may occur in the phase data . The removal of these steps, called phase unwrapping, is necessary to make the phase data continuous. Due to speckle decorrelation[7,8] and noise the detection of 2π-steps in phase data obtained by speckle interferometry is not straightforward. The most popular and straightforward way to unwrap the phase data is shown in Fig. 6. This method is based on adding an offset (a multiple of 2π) to the phase value at each pixel and consists of the following passes. Starting at the top left pixel of the image, the offset is set to zero. Then the first column is scanned to determine the offsets of the first pixel of each row. The offset changes each time a 2π-step is detected. Finally for each row the offsets are calculated starting with the offset in the first column. A 2π-step is detected if the absolute difference between a pixel and the previous pixel exceeds π. It is usually more convenient to locate the starting point at the centre of the image.

The scanning method mentioned above is successfully used in classical interferometry. For speckle interferometry this method is not suitable, because the discontinuities can be hidden in speckle noise, causing steps to be neglected or offsets to be added at the wrong points. A further restriction of the algorithm mentioned above is that all pixels in the input image have to be valid data pixels. If there are invalid pixels, i.e. B_m is not empty, these pixels have to be corrected before phase unwrapping. This is called phase restoration and is done by substituting the value of a valid pixel from the $3*3$ neighbourhood of the invalid pixel. Averaging neighbourhood pixels is not possible because of the possibility that 2π-steps

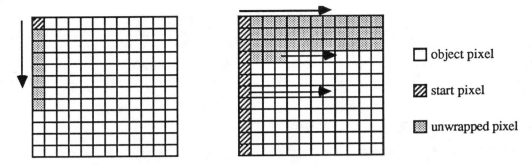

Fig. 6. Phase unwrapping by scanning each row after scanning the first column.

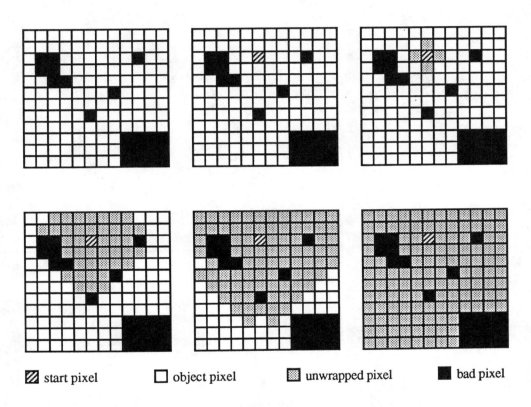

start pixel   object pixel   unwrapped pixel   bad pixel

Fig. 7. A phase unwrapping algoritm that scans the data using a pixel queue.
Different stages are shown from choosing a start pixel to the unwrapped result.

are present. If no valid pixel is available in the neighbourhood, the pixel remains an invalid pixel. More iterations would correct that pixel, but then the systematic error in the output data increases.

The computation of a continuous phase map can be improved by excluding the invalid pixels from the unwrap procedure. We have developed an algorithm based on a pixel queue[9] to scan the data like a fluid flowing over the object around invalid pixels (Fig. 7). Pixel addresses are stored on one side of the queue and fetched from the other side. The following steps are involved:

1.   A pixel somewhere in the source image is chosen as start pixel.
2.   If the start pixel is not a valid pixel, the 3*3 neighbourhood is searched for a valid one.

3. The addresses of the non-processed neighbours of that pixel are stored on a queue.
4. An address is taken from the queue and the corresponding pixel will be processed. If the queue is empty go to 10.
5. If this pixel is a valid pixel, the pixel value is compared to the mean of the unwrapped, valid pixels in a certain neighbourhood. Else go to step 3.
6. If a step is detected, the current pixel is corrected by adding or subtracting a multiple of 2π.
7. Goto step 3.
8. If the valid data is not completely unwrapped, 1 to 7 can be repeated using another start pixel.

Of course this can lead to discontinuities between processed areas.

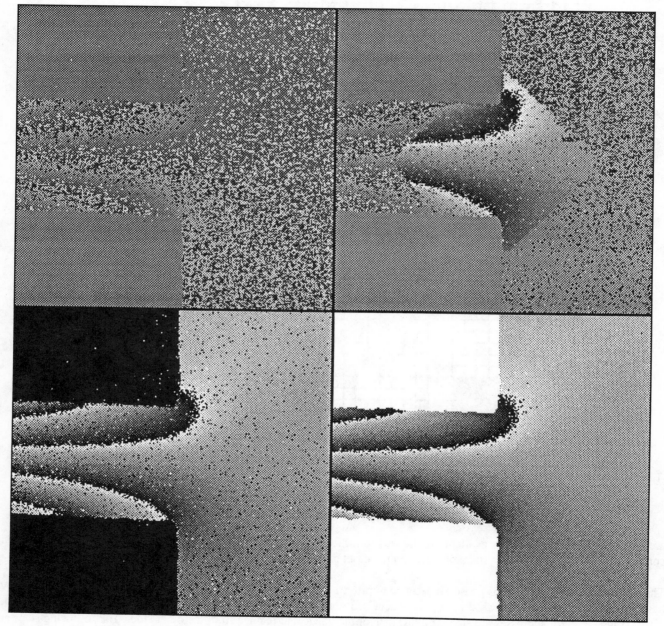

Fig. 8. Measured phase change ϕ modulo 2π, when the object is loaded (a).
Phase unwrapping. The algorithm propagates a "wavefront", neglecting invalid pixels (b).
Unwrapped phase change in the x-direction. Continuous phase map (c).
Restored phase map: invalid pixels have been replaced by the average of their valid neighbours (d).

In this case 4-connectivity is used. This means that the neighbours of a pixel are those pixels that share an edge with that pixel. With 4-connectivity the processing is propagating in a diamond shape. If also the 8-connected neighbours are taken into account (pixels that share a corner with the central pixel) a square wavefront occurs. Because errors are more likely at the corners of a wavefront, new algorithms are developed now using pixel buckets, leading to a circular wavefront. In step 5 a 3*3 neighbourhood is taken to compute the mean.

The phase unwrapping algorithm mentioned above allows invalid pixels, caused by low modulation or saturation, to occur in the input image. After phase unwrapping, these pixels are replaced by the average of their valid neighbours in a 3*3 neighbourhood. A number of iterations can be performed to restore the data, but the systematic error in the output data increases after each iteration. The phase restoration is not really necessary for further processing steps. However, it creates a better looking result.

To produce the binary image that marks the background, the image B_m is processed using the following basic binary image processing techniques[9,10]:

- erosion (thinning), i.e. changing all object pixels connected to the background into background pixels.
- dilation (expanding), i.e. changing all background pixels connected to the object into object pixels.
- opening, i.e. an erosion followed by a dilation.
- fill, i.e. changing all background pixels lying in a closed contour of object pixels into object pixels (fill small not unwrapped holes).

With the fill operation areas within closed contours of B_m are filled. This is done before a phase restoration step. After each phase restoration step an opening is performed on B_m to produce the corresponding mask. In this case, for all operations 4-connectivity is used. Finally, the result image B_m contains a mask that separates background from object. Fig. 8 shows an intermediate and the final result of the phase unwrapping algorithm and also the result of the phase restoration. The discontinuities in Fig. 8 are due to the display range (after phase unwrapping the greyvalues exceed the 0-255 range). In The data in computer memory is continuous. Rescaling can be used to force the greyvalues into the 0-255 range for display.

6. DETERMINATION OF DISPLACEMENT AND STRAIN COMPONENTS

For each point on the surface of the object the measured phase change ϕ is directly related to the displacement vector **L** by:

$$\phi = \mathbf{K \cdot L} = (2\pi/\lambda)\,(\hat{i}_i + \hat{i}_v) \cdot \mathbf{L}, \tag{4}$$

where **K** is the sensitivity vector, $\hat{i}_i$ and $\hat{i}_v$ are unit vectors in the illumination and viewing direction and λ is the wavelength[3]. In general, the sensitivity vector depends on the position at the surface of the object and thus on the position in the image that contains the phase data. In the case of in-plane displacement measurements the calculations can be simplified considerably by performing *four* phase measurements with collimated illuminating beams to cancel the dependence of the illumination direction $\hat{i}_i$ on the position at the object surface. We measure the phase changes ϕ_1 and ϕ_3 with illumination directions lying in the x-z-plane and the phase changes ϕ_2 and ϕ_4 with illumination directions lying in the y-z-plane. Both pairs of measurements are performed with two illumination directions that are symmetrical with respect to the z axis. By subtracting the measurement results for opposite illumination directions, the in-plane displacement components can easily be calculated, because the resulting sensitivity vectors will be along the x and y axes.

$$L_x = [\phi_3 - \phi_1]\,/\,[(2\pi/\lambda)2\sin\alpha] \tag{5}$$
$$L_y = [\phi_4 - \phi_2]\,/\,[(2\pi/\lambda)2\sin\alpha], \tag{6}$$

where α is the angle between the illumination direction $\hat{i}_i$ and the z axis[11,12].

The in-plane strain components are the normal strains e_{xx} and e_{yy} and the shear strains e_{xy} and e_{yx}. For a flat object surface normal to the z axis they are defined as:

$$e_{xx} = \partial L_x/\partial x \qquad e_{yy} = \partial L_y/\partial y \qquad e_{xy} = e_{yx} = \{\, \partial L_x/\partial y + \partial L_y/\partial x \,\}\,/2. \tag{7}$$

The calculation of these strain components involves the determination of the first derivative of the in-plane displacement components. We used a linear least squares fit of a plane (LLS) to a set of valid neighbouring data points for each data point, supplying the derivative with respect to x and y at the same time[5]. This enabled us to measure the in-plane strain accurately near the object edges too. As will be shown below, this method can also be applied successfully as a smoothing filter.

For each point (x,y) in the image a rectangular neighbourhood is considered (Fig. 9). This neighbourhood contains a set of N(x,y) valid data points (m,n,g(m,n)), where (m,n) is the position of a data point and g(m,n) is the greyvalue at that point. The position (m,n) is relative to the neighbourhood ((0,0) is assumed to be at the centre of the neighbourhood). A LLS fit of a function f(m,n) at location (x,y) can be performed by minimizing the following expression:

$$S(x,y) = \sum_{\forall\ (m,n)} (f(m,n) - g(m,n))^2. \tag{8}$$

The summation $\sum$ is done over all the valid data points (m,n) in the rectangular neighbourhood. Substituting the general expression for a plane in the three-dimensional space for f(m,n) gives:

$$S(x,y) = \sum (a*m + b*n + c - g(m,n))^2. \tag{9}$$

Minimizing Eq. 9 is done by solving the following set of equations:

$$\frac{\partial S}{\partial a} = 0 \qquad \frac{\partial S}{\partial b} = 0 \qquad \frac{\partial S}{\partial c} = 0 \tag{10}$$

or, substituting Eq. 9 :

$$
\begin{aligned}
a \sum m*m &+ b \sum n*m &+ c \sum m &= \sum m*g(m,n) \\
a \sum m*n &+ b \sum n*n &+ c \sum n &= \sum n*g(m,n) \\
a \sum m &+ b \sum n &+ c N &= \sum g(m,n).
\end{aligned}
\tag{11}
$$

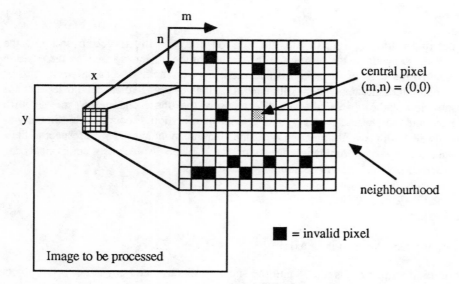

Fig. 9. Fitting a plane in a rectangular neighbourhood of (x,y) to estimate the first partial derivatives of the data.

This matrix equation is solved for each position (x,y). If no invalid pixels are present in the neighbourhood, Eq. 11 reduces to:

$$
\begin{aligned}
a \sum m*m &= \sum m*g(m,n) \\
b \sum n*n &= \sum m*g(m,n) \\
c\,N &= \sum g(m,n).
\end{aligned}
\qquad (12)
$$

In that case a, b and c are easily derived from Eq. 12. If invalid pixels are present, a matrix inversion is used to compute a, b and c from Eq. 11. Because a variable amount of invalid data points can occur in the neighbourhood, the solution of

a	b	c	d
e	f		i
g	h		

Application of the linear least square fit (LLS) to smooth the data and to compute the first derivatives of the data to get the in-plane strain components.

Fig. 10. The in-plane phase change in the x-direction (a), the in-plane displacement in the x-direction (b) and the y-direction (c) with a contour spacing of 340 nm after a 25*25 smoothing filter, the in-plane normal strain in horizontal direction $\partial L_x/\partial x$ (d), the in-plane normal strain in vertical direction $\partial L_y/\partial y$ (e), the derivatives $\partial L_x/\partial y$ (f) and $\partial L_y/\partial x$ (g) and the in-plane shear strain component (h) calculated from image f and g. Image i shows the iso-strain lines of image h with a contour spacing of 3 μstrain.

Eq. 11 is not straightforward. All the summations have to be calculated each time the window (neighbourhood) shifts one pixel to the right. To reduce the computation time, the summations in that window are computed using a special updating mechanism. First the window is shifted to the right, starting at the topleft of the image, till at least one valid data point is located in the window. Then for that window all summations and the number of valid data points N are calculated. For the computation of $\sum m*m$, $\sum n*n$ and $\sum n*m$ a table-lookup is used. After that, the window is shifted one pixel to the right and the number of invalid pixels N and the summations are updated using the values in the added column on the right and the removed column on the left. Because the set of invalid pixels in the un-updated part of the window changes after a shift to the right, the following corrections are necessary *after* updating:

$$
\begin{aligned}
(\textstyle\sum m*m)' &= (\textstyle\sum m*m) - 2*(\textstyle\sum m) + N \\
(\textstyle\sum m*n)' &= (\textstyle\sum m*n) - (\textstyle\sum n) \\
(\textstyle\sum m*g)' &= (\textstyle\sum m*g) - (\textstyle\sum g) \\
(\textstyle\sum m)' &= (\textstyle\sum m) \qquad - N,
\end{aligned} \tag{13}
$$

where the primed values denote the corrected values for the previouly computed summations. The corrections have to be performed in the above order.

For each line in the image the above procedure is repeated. The values a(x,y) and b(x,y) are an estimation of the first derivative with respect to x and y. Near the object edges a systematic error can occur. This error can be reduced by performing a second order LLS fit, but then the noise reduction will be less. The computation of c(x,y) provides a smoothing filter, that is identical to a uniform convolution filter if no invalid pixels are present, but performs much better near object edges because the result is hardly influenced by the occurrence of invalid pixels. Fig. 10 shows the result of the the LLS fit used as smoothing filter and as gradient filter. Also the results of the computation of in-plane strain components is shown in Fig. 10.

7. ACCURACY

In our present experimental system the accuracy of a phase change measurement is mainly determined by electronic noise, speckle decorrelation, systematic errors and changes in the environmental conditions. The electronic noise source is the CCD camera, having a measured r.m.s. signal-to-noise ratio of approximately 100. The phase error caused by speckle decorrelation is randomly distributed over the speckles and can therefore be reduced by applying a smoothing filter as discussed in paragraph 6. The nature of a difference measurement causes most systematic errors to become less significant. A systematic error like the determination of the sensitivity vector from the geometry of the interferometer however, will effect the result directly. Particularly in a non-laboratory environment, vibrations and air turbulence can cause noise and apparent systematic errors, making vibration isolation and thermal shielding of the optical set-up necessary. In our laboratory system, that digitizes four phase shifted interference patterns in 160 ms, the measurements could be accurately performed without isolation and shielding.

The application of the smoothing filter, described in paragraph 6, gives a noise reduction proportional to the filter size. The spatial resolution is approximately inversely proportional to the filter size. At edges of the object the accuracy decreases, because less pixels are taken into account. The repeatability of the measured strain at each pixel, when using a 45*45 pixel filter, amounts to approximately 0.3 μstrain r.m.s.. For small loads the estimated inaccuracy approaches the repeatability, because the speckle decorrelation becomes negligible. For the measurement presented in this paper an area of 3.5 mm^2 around each object point (corresponding to 45*45 pixels) was used to calculate the derivatives.

The upper limit of the measuring range of the strain depends on the size of the object region that is imaged on the detector and the maximum phase fringe density that can still be successfully unwrapped. For a 100 x 100 mm^2 object region the upper limit will be circa 200 μstrain.

8. CONCLUSIONS

A measuring method to determine in-plane displacement components and the in-plane strain components of a flat surface of a deformed object has been presented. Application of a special gradient filter enables accurate determination of the in-plane strain components at each point of the object surface. Special purpose hardware allows a complete measurement of the phase change modulo 2π on a 512*512 grid every 240 ms. We have developed new image processing algorithms for phase shifted interference patterns, such as phase unwrapping, phase restoration and smoothing, that have proven to be very robust with respect to both temporal and decorrelation noise, the modulation amplitude and object shape. It takes about 2 minutes to determine a continuous 512*512 phase map, including measurement of the reference phase data, on the VME/MC68000 system. The processing time of the LLS fit strongly depends on the filter size. At the moment this calculation is done on a SUN3 workstation and takes about 8 minutes when a neighbourhood of 45*45 pixels is used. The ability to distinguish the object from the background, has greatly improved the flexibility of the processing of phase stepped interference patterns. The accuracy of the result of our strain measurement using a local fitting technique indicates that this method is a serious alternative for strain gauges.

Recently new speckle interferometric techniques have been reported[13,14], that diminish the influence of environmental disturbances on the phase measurement or enable the analysis of transient events. In particular for measurements in a non-laboratory environment a combination of these techniques and the techniques presented in this paper has potential to become an important measuring method.

9. ACKNOWLEDGEMENT

The presented work has been performed under sponsorship of the "Stichting voor Fundamenteel Onderzoek der Materie" under project nr. DTN98.0329.

10. REFERENCES

1. K. Creath, "Phase shifting speckle interferometry", Applied Optics **24**, p. 3053, 1985.
2. D.W. Robinson and D.C. Williams, "Digital phase stepping speckle interferometry", Optics Comm. **57**, p. 26, 1986.
3. C.M. Vest, *Holographic interferometry*, John Wiley & Sons, New York, 1979.
4. A.A.M. Maas and H.A. Vrooman, "Digital phase stepping speckle interferometry", *Laser Technologies in Industry*, O.D.D. Soares ed., Proc. SPIE 952, p. 196, 1989.
5. H.A. Vrooman and A.A.M. Maas, "Image processing in digital speckle interferometry", *Proc. of the 1989 Fringe Analysis conference*, Loughborough Consultants Ltd., Loughborough, 1989.
6. D.W. Robinson, personal communication.
7. A.E. Ennos, "Speckle interferometry", in *Laser speckle and related phenomena*, J.C. Dainty ed., chapter 6, Springer-Verlag, New York, 1975.
8. R. Jones and C. Wykes, "De-correlation effects in speckle-pattern interferometry / 2. Displacement dependent de-correlation and applications to the observation of machine induced strain", Optica Acta **24**, p. 533, 1977.
9. L.J. van Vliet and B.J.H. Verwer, "A contour processing method for fast binary neighbourhood operations", Pattern Recognition Letters **7**, p. 27, 1988.
10. A. Rosenfeld and A. Kak, Digital Picture Processing, Vol. II, Chapter 11, Academic Press, Orlando, 1982.
11. A.A.M. Maas and H.A. Vrooman, "Strain Measurement by Digital Speckle Interferometry", in *Proc. of 12th World Conference on Non Destructive Testing*, J. Boogaard and G.M. van Dijk eds., Elsevier Science Publishers B.V. , Amsterdam, 1989.
12. A.A.M. Maas and H.A. Vrooman, "In-plane strain measurement by digital phase shifting speckle interferometry", *Fringe Analysis*, Proc. SPIE 1162, to be published.
13. M. Kujawinska and D.W. Robinson, "Multichannel phase-stepped holographic interferometry", Applied Optics **27**, pp. 312-320, 1988.
14. F. Mendoza Santoyo, D. Kerr, J.R. Tyrer, "Interferometric fringe analysis using a single phase step technique", Applied Optics **27**, pp. 4362-4364, 1988.
15. TCL-Image User's manual, reference number TCLI-88-005, Multihouse TSI b.v. technical, scientific and industrial systems, Copyright TNO Institute of Applied Physics, Delft, The Netherlands, 1988.

SESSION 2

Image Processing Techniques

Chair
Kevin G. Harding
Industrial Technology Institute

FRINGE LOCATION BY MEANS OF A ZERO CROSSING ALGORITHM

K.J. Gåsvik, *), K.G. Robbersmyr, **) and T. Vadseth ***)

*)On leave from SINTEF, Division of Machine Design
**)Division of Machine Design
***)SINTEF, Division of Safety and Reliability

The University of Trondheim
The Norwegian Institute of Technology
N-7034 Trondheim, Norway

ABSTRACT

Moiré technique using projected fringes is a suitable method for full field measurements of out-of plane deformations and object contouring. In many applications however, the sensitivity of the method is often too low. To overcome this problem, one solution is to apply interpolation techniques by means of precise location of the fringe maxima.

In this paper we present a method of fringe location by means of a zero crossing algorithm. The system is based on a digital image processing module plugged into a PC/AT using a CCD camera as the imaging sensor. In this way, the positions of the maxima are located with an accuracy of a fraction of one pixel width.

1. INTRODUCTION

The increasing number and capacity of digital image processing plug-in modules for personal computers has renewed the interest in moiré topography. In combination with electronic imaging devices, these modules offer a convenient means for image processing and fringe analysis.

In this paper a system based on moiré technique with projected fringes is presented. Unlike most of similar methods reported in the literature 1-5, this system does not employ a reference grating in front of the imaging sensor 6-11. It is therefore the resolving power of the imaging sensor that limits the sensitivity of the method.

To increase the sensitivity, a zero crossing algorithm for precise location of the fringe maxima is developed. A detailed description of this algorithm is presented together with a simple experiment which shows the optimum fringe period (in pixels) to be employed with the algorithm.

2. SYSTEM DESCRIPTION

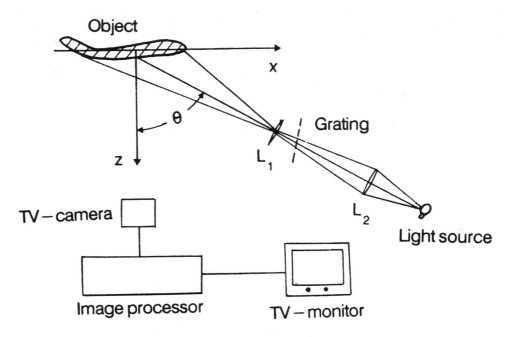

Figure 1 Experimental set up

Fig. 1 illustrates the experimental set up. A grating is projected onto the object surface under an angle θ to the z-axis by a projection lens L_1. The projection unit consists of a mercury high pressure arc lamp, a condensor lens, a grating holder and a photographic objective as the projection lens. A CCD Camera images the object along the z-axis and the video signal is input to an image processor card plugged into a PC/AT.

The common procedure to obtain a map of the outward deformation of an object is: 1) Record and store an image of the object with projected fringes before deformation and 2) Subtract this image from the image of the object after deformation. The resulting image superposed with moiré fringes can be displayed on a TV monitor. These moiré fringes map the surface deformation with a contour interval given by

$$\Delta z = d/\sin\theta \tag{1}$$

with

$$d = md_g \tag{2}$$

where d_g is the grating period and m is the magnification of the projection system. The angle θ will vary across the object but an exact formula for θ as a function of the x coordinate can be obtained from one law of sines in a triangle.

The same procedure can also be applied to obtain a map of the surface height difference between two different objects or an object and a computer generated master.

3. INTERPOLATION TECHNIQUE

For surface height deviations greater than the contour interval given by eq. (1), a map of moiré fringes is usually easily obtained applying the described procedure. Due to the low resolution of the TV camera however, this contour interval is limited. It is therefore often desireable to be able to measure smaller deviations than this contour interval. One can then apply the following interpolation technique.

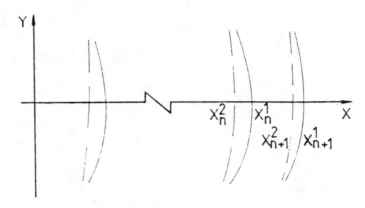

Figure 2 Positions of the projected fringes on object 1
(solid lines) and object 2 (dashed lines)

In fig. 2, assume that the solid lines and the dashed lines are the projected fringes on object 1 and object 2, respectively. Object 1 and 2 might be the same object before and after deformation, or a master and a product specimen. Along a line parallel to the x-axis, the positions of the n-th projected lines are denoted x_n^1 and x_n^2 for object 1 and object 2 respectively, and the fringe periods are

$$d_n^1 = x_{n+1}^1 - x_n^1 \qquad\qquad (3a)$$

$$d_n^2 = x_{n+1}^2 - x_n^2 \qquad\qquad (3b)$$

Note that nowhere is the difference

$$\triangle x_n = x_n^2 - x_n^1 \qquad\qquad (4)$$

larger than d_n^1, i.e. we have assumed a surface height deviation smaller than the contour interval.

Now, at each position x_n^2 the surface height deviation $z(x_n^2)$ can be calculated as

$$z(x_n^2) = \frac{\triangle x_n}{d_n^1} \cdot \triangle z \qquad\qquad (5)$$

where $\triangle z$ is given from eq. (1). The value of the deviation z between the projected lines is found by calculating the straigt line connecting $z(x_n^2)$ and $z(x_{n+1}^2)$ (linear interpolation). The validity of this interpolation is of course dependent of the curvature of the deviation relative to the fringe period d.

4. ZERO CROSSING ALGORITHM

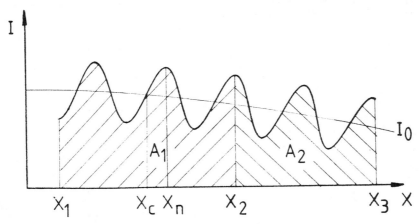

Figure 3 Intensity distribution I of the projected fringes
with mean intensity I_0

In fig. 3, the intensity I of the projected fringes along a TV-line (the x-direction) is plotted. Each pixel element i has an intensity value equal to I_i. The problem is to localize the positions x_n of the fringe maxima as exact as possible. Because of the signal noise, this has shown to be difficult by using e.g. some kind of moment of inertia calculations. Instead we try to localize the positions x_c where the intensity crosses the mean intensity line I_0. Since the lightening conditions vary, this mean intensity value is normally not constant. Here we approximate this value by a straight line given by

$$I_0 = ax + b \qquad (6)$$

Assume that the TV-line is scanned from pixel no $i = x_1$ to $i = x_3$. We divide this interval into two equal subintervals at pixel no. $x_2 = (x_3 + x_1)/2$ and where $N = x_3 - x_1$ is the total number of pixels along the scanned line. Then the constants in eq. (6) is found to be given by

$$a = \frac{4(A_2 - A_1)}{N^2} \qquad (7a)$$

$$b = \frac{3A_1 - A_2}{N} - \frac{4(A_2 - A_1)}{N^2} x_1 \qquad (7b)$$

Here A_1 and A_2 are the areas under the line given by eq. (6) from x_1 to x_2 and from x_2 to x_3 respectively. As can be seen, b is dependent on x_1, i.e. the origin of the coordinate system.

If the intensity distribution of the projected fringes behaves properly, theese areas will also be given by

$$A_1 = \sum_{i=x_1}^{x_2} I_i \qquad (8a)$$

$$A_2 = \sum_{i=x_2+1}^{x_3} I_i \qquad (8b)$$

In this way the mean intensity given by eq. (6) is found by summing the intensity values of each pixel along the TV-line using eqs. (7) and (8).

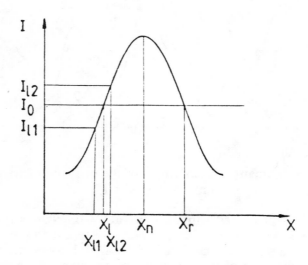

Figure 4 Detection of the n-th maximum by locating
the left (x_l) and right (x_r) crossover points

By moving from left to right, the position x_n of the n-th projected fringe is found by the following procedure: (see fig. 4) 1) The last pixel with intensity value I_{l1} below I_o and 2) the first pixel with intensity I_{l2} over I_o are detected. 3) The straigt line connecting I_{l1} and I_{l2} is determined, and 4) The intersection between this line and the mean intensity line is found, giving the cross-over position x_l. The same procedure is repeated to locate the right crossover point x_r and the position of the intensity maximum of the fringe is given as

$$x_n = (x_l + x_r)/2 \qquad (9)$$

This procedure of locating the fringe positions can of course be improved in many ways, but has proven to work properly.

The result of this algorithm is that we get the value of the surface height deviation at each pixel in the image by means of eq. (5). This result can be displayed in many ways. Fig. 5 shows the deviation of a propeller blade from a computer generated master displayed as a greytone picture.

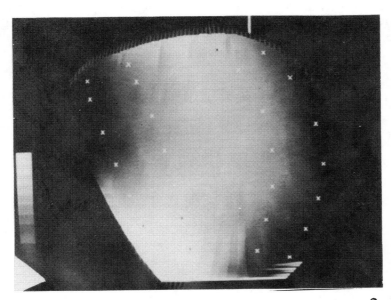

Figure 5 Deviation of a real propeller blade (1 x 0,9 m^2)
from a computer generated master displayed as a greystone picture.
Dark areas = negative deviations, light areas = positive deviations.

5. EXPERIMENTS

An important aspect of the system is the lower limit of the measureable surface height deviations.
This will depend strongly on how accurate the positions of the projected fringes can be detected.
With the above described algorithm, this again will depend on a lot of parameters, among others,
the camera, its gamma factor and especially the jitter of the line sync signal. But above all, it
will depend on the object surface which should preferably be uniformly white and diffusely reflecting.

To see how the zero crossing algorithm works with varying fringe periods, a simple experiment was
undertaken. The object was a rectangular flat plane (10 x 10 cm^2) which could be rotated about an
axis parallel to the x-axis. We used a grating with 40 lines per mm (d_g = 0,025 mm) a magnification
m = 24 and varied the fringe period by varying the projection angle θ. The plane was given a
controllable rotation by means of a micrometer screw between the two recordings. As a measure
of the accuracy of the fringe location, the mean $\triangle x$ and standard deviation $\sigma \triangle x$ from about
300 pixels in the image were taken. The result is plotted (the dashed curve) as a function of the
fringe period d_x in fig. 6. From this we see that the absolute positions of the fringes are located
with highest accuracy (less than one tenth of a pixel) for a fringe period of about 6 pixels. For
fringe periods lower than about 4 pixels, the algorithm had problems in locating the fringe maximum
at all, a fact that can be realized from the sampling theorem. What is most important however, is
the relative accuracy of the fringe location,   i.e. the quantity $\sigma \triangle x/d_x$. This is plotted versus
d_x in fig. 6. From this it is seen that the relative accuracy improves with increasing fringe period
up to a period of about 10 pixels, whereafter it is virtually constant. With d_x = 10 pixels we see
that the relative positions of the fringes are detected with an accuracy of 1 per cent, indicating
that the contour interval $\triangle z$ can be improved by a factor of 100. The experiment also shows that
it is not necessary or even preferably to increase the fringe period above 10 pixels to improve the
accuracy.

In our set up, with $\triangle z = 0{,}6$ mm (for $\theta = 80^\circ$), it was therefore possible to measure the surface height deviation with an accuracy of 6 µm. It should however be remembered that the object was a flat plane and the experimental conditions the best possible.

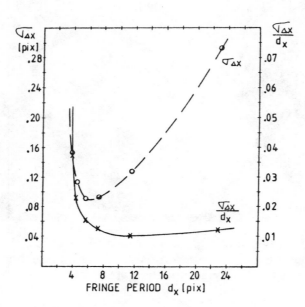

Figure 6 Standard deviation $\sigma\triangle x$ (dashed curve) and relative standard deviation $\sigma\triangle x/d_x$ (solid curve) of the fringe positioning plotted against fringe period d_x.

6. REFERENCES

1. G.T. Reid, R.C. Rixon and H.I. Messer, "Absolute and comparative measurements of three-dimensional shape by phase measuring topography", Optics and Laser Technology 16, 315-319 (1984).
2. M. Idesawa, T. Yatagai, and T. Soma, "Scanning moiré method and automatic measurement of 3-D shapes", Appl. Opt. 16, 2152-2162 (1977).
3. J.C. Perrin and A. Thomas, "Electronic processing of moiré fringes: Application to moiré topography and comparison with photogrammetry", Appl. Opt. 18, 563-574 (1979).
4. L. Pirodda, "Shadow and projection moiré techniques for absolute or relative mapping of surface shapes", Opt. Eng. 21, 640-649 (1982).
5. M. Hhalioua, R.S. Krishnamurthy, H. Liu and F.P. Chiang, "Projection moiré with moving gratings for automated 3-D topograpy", Appl. Opt. 22, 850-855 (1983).
6. T. Yatagai and M. Idesawa, "Automatic fringe analysis for moiré topograpy", Optics and Lasers in Engineering 3, 73-83 (1982).
7. K.J. Gåsvik, "Moiré technique by means of digital image processing", Appl. Opt. 22, 3543-3548 (1983).
8. K.J. Gåsvik, "Moiré contouring of objects generated by CAD", Proc. SPIE vol. 814, Photomechanics and Speckle Metrology, Ed.: F.P. Ciang, 1987.
9. K.J. Gåsvik, T. Hovde and T. Vadseth, "Improving industrial inspection by advanced mapping technology", Presented at The 5th CIM-Europe Conference, Atens, 16-18 May 1989.
10. K.J. Gåsvik, T. Hovde and T. Vadseth, "Moiré technique in 3-D machine vision", To be publised in Optics and Lasers in Engineering 1989.
11. K.J. Gåsvik, "Optical metrology", pp. 125-144, John Wiley & Sons, Cichester, 1987.

New Method of Extracting Fringe Curves from Images

K. Liu and J. Y. Yang

East China Institute of Technology
Department of Computer Science
Nanjing Jiangsu, China

ABSTRACT

The extraction of fringe curves is the foundation of the automatic quantitative analysis of interferometric images. This paper proposes an efficient method for interferometric fringe curves extracting. First, all fringe curve points are detected. Then these fringe curve points are tracked. Finaly, the fringe curves are extracted by a linking algorithm. The detection principal of the fringe curve points and criterions for curve point tracking and interrupted gaps linking are discussed in detail in the paper. The experimental results show that the method is effective.

1. INTRODUCTION

The quantitative analysis of interferometric fringe images has wide applications in optical parts measurement, 3-D information acquisition and nondestructive testing[1-2]. The tranditional processing mean by hand is tedious and the results have low precision. The way by using computer digital image processing techniques can not only increase the precision and calculation velocity but also save the manpower. The key problem of the automatic quantitative analysis of interferometric fringe images is extracting fringe curves from an image. Although several kinds of methods[3-7] have been presented for fringe curve extraction, the problem, however, is still unresolved due to the noise in fringe images. The method presented in the paper is simple and can be used in noisy case. The implementation of the algorithms can be divided into two steps. In the first step, all fringe curve points are found out in an image, then the fringe curves are extracted by a tracking and linking algorithm. The fringe curve detection algorithm is designed based on the width curve detection principal presented in paper[8]. In all two steps, a priori information about fringe images has been used. The experimental results show that our method is succesful.

2. FRINGE CURVE POINT DETECTION

In paper[8], we have proposed a width curve detection method that can effectively detect the curve points from an image which involves a complex background and a lot of noise. Since our fringe curve point detection algorithm is derived from the principal of the above method, so we first introduce the width curve detection principal in the following.

2.1. Width curve detection principal

Let the curve which is smooth and continuous have width W. This width curve can be considered as a thin and long region in an image plane(called curve region). Let C1 and C2 be its two long boundaries(see Fig. 1). Since the curve region has smooth width, hence C1 and C2 have same shape and C1 is approximately parallel to C2. For any point P on C1, along its inner normal direction, there is a point P on C2 such that

a. the distance between P and P' is W
b. the difference of the inner normal directions between P and P' is 180 degree.

P' is called the corresponding edge point of P.

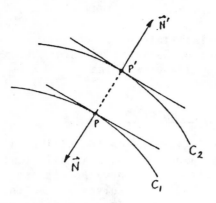

Fig. 1. Smooth curve region

A searching algorithm has been designed in the paper based on above two characteristics. For any curve region edge point, search out its corresponding edge point, the median point of the two points is taken as curve point.

2.2. Fringe curve detecting algorithm

Assume that the fringe curve points we want to detected are ' dark ' fringe curve points. A fringe in an interferometric images can be considered as a width curve from the view of the field of digital image processing, so we use above width curve detection principal to extract fringe curves. The fringe curve points have the smallest values of grey levels along their normals in the local areas. This feature can be combined with above principal to detect fringe curve points. In the following, we will describe the basic steps of the fringe curve detection algorithm. In order to suppress the noise in an image, a ponit-like noisy smooth and filter operator is used in the algorithm.

2.2.1. Smoothing the isolated points

There are two types of noise: isolated points and undesired small regions. We use a local average smoothing operator to decrease the effect of isolated noisy point.

In Fig. 2, a_{ij} (i, j=1,2,3) is the point's grey values. The Average grey value for a_{22} calculated by the following formula

$$\bar{g} = \frac{a_{22}}{4} + \frac{a_{21} + a_{12} + a_{23} + a_{32}}{8} + \frac{a_{11} + a_{13} + a_{31} + a_{33}}{16}$$

a_{11}	a_{12}	a_{13}
a_{21}	a_{22}	a_{23}
a_{31}	a_{32}	a_{33}

Fig. 2. 3×3 window.

2.2.2. Edge point detection and its inner normal directiion calculation

In order to detect fringe curve points, we must first find out the edge points of all fringe curve regions and give their inner normal directions. We use Sobel operator to detect the edge points of curve regions, because it is simple and can give the grey gradient directions of the edge points. After the original image has been processed by the average smoothing operator, the image is processed by Sobel operator. Then the edge points can be selected by the techniques of thresholding and non-maximun values suppresion techniques.

The inner directions of the edge points of the curve regions can be calculated by modifying the grey gradient directions. Readers may referee to the paper for the details about the detection of the edge points and the calculation of their inner normal directions' formulas.

2.2.3. Remove undesired small regions

The small regions which consist of undesired points can't be removed by average smoothing operator. After above two steps, edge points of these regions have the following features: edges are close curves which have shorter perimeters, or the distance between two end points of the curve segements is very small.

If the difference of normal directions between two consecutive candidate edge points is smaller than the tolerant limit, the two points are called connected edge points. The sets composed of all connective edge points are called connective sets of edge points. The length of a connective set S is defined as follows

$$L = max (\; max \; (\; |x_i - x_j|, \; |y_i - y_j|) \;)$$
$$\begin{array}{c}(x_i, \; y_i) \\ (x_j, \; y_j)\end{array} \in S$$

L can be used to suppres the shorter edge curve segements or close edge curves which have small perimeters.

2.2.4. Detecting fringe curve points

After above three steps, the fringe curve points can be detected by the following method.

For any edge point P, search for its corresponding point P' along its inner normal direction. Then, the point which has the smallest grey value among the points lying on the segement $\overline{PP'}$ in original image is selected as fringe curve point. In order to remove the noise' effect, we may average the points' grey values of the points on PP before selecting the fringe curve points.

3. FRINGE CURVES TRACKING AND INTERRUPTED GAPS LIKINNG

After the original image is processed by the fringe curve detection algorithm, the fringe curve points in the image are obtained. However, there are many interrupted gaps of the fringe curves due to the noise and the error of the formula for calculating the directions of the grey gradients. We have designed a fringe curve tracking and the interrupted gaps linking algorithm based on the following characteristics: fringe curves are continuous and two different fringe curves do not intersect each other. The follows are the steps of our algorithm.

3.1. Tracking the continuous fringe curve segements

In the first step, we use a tracking algorithm to select all connective fringe curve segements in which the differences of two consecutive curve points are within the tolerant limit. For every fringe curve segement tracked out, calculate the tangent directions of its initial and final point. Here, the tangent direction of a initial point is opposite to the tracking direction and the one of a final point is consistent with tracking direction. All initial points and final points are called interrupted points of the fringe curve segements.

3.2. Identifying small interrupted gaps and interpolating

After above step , for any interrupted point, search for other interrupted point in a sectorial area determined by the parameters β and W as shown in Fig. 3. Here, W is only taken few number of pixels' width. If there is an interrupted point in the sectorial area, these two points are linked by the points calculated by a two order polynomial interpolating based on the border conditions: the positions of the two points and their tangent directions.

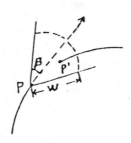

Fig. 3.   Sectorial search area

3.3. Identifying larger interrupted gaps and interpolating

After above two steps, there are still some interrupted gaps which are not interpolated. In general case, they correspond to those undesired small regions. Since a fringe curve is continuous and different fringe curves do not intersect each other, for points P_1 and P_2, if they are two interrupted points of an interrupted gap, the following conditions must satisfy

(1) segement $\overline{P_1P_2}$ do not intersect other fringe curves.
(2) for two interrupted points P_3, P_4 which form an interrupted gap, $\overline{P_1P_2}$ do not intersect $\overline{P_3P_4}$.

Therefore, we may design an algorithm for larger interrupted gaps identifying and interpolating based on above two characteristics. The following is the ideal of the algorithm.

For any interrupted point P, find out all other interrupted points in the sectorial area as shown in Fig. 3. Here, W is taken larger than the one in case of above step and the number of the interrupted points here may be larger than 1. All the interrupted points are called candidate corresponding points of P. The set consists of these points is called candidate set. If P' is a candidate corresponding point and P ,P' are two interrupted points of an interrupted gap, then P' is called true corresponding point of P. When there are candidate corresponding points in the sectorial, order all these points based on their rights which are inverse proportion to their distances from P and the deferences between their tangent directions and P's tangent direction. After then, the following procedure is used to find true corresponding point of P.

Select point P_1 from the candidate set according the rights of the points in the candidate set to inspect if the following conditions are satisfied

(1) there is no any curve points on $\overrightarrow{PP_1}$.
(2) if there are two other candidate corresponding points P_2 and P_3 satisfying :
   (i). P_2 is in P_3's searching sectorial area and P_3 is also in P_2's searching sectorial area , here the sectorial areas is the same as the P's; (ii). P_2 and P_3 have not been inspected.

If there is P_1 satisfing above conditions, P_1 is taken as true corresponding point of P and the inspecting procedure ends.
The interpolating method here is the same as the one in the case of above step.

4. EXPERIMENTAL RESULTS

We have developed an algorithm on PCVISION image processing system based on the algorithms as stated in this paper. A lot of experiments have been done for the practical interferometric fringe images. Experimental results show that our method can extract fringe curves which have good position precision. In the following, we will give some processed results about an image to show the main implementation steps of the algorithms.
Fig. 4 illustrates an original laser interferometric fringe image. There is a lot of noise in the image due to the various effects. Fig. 5 is the result processed by the edge detection and inner normal direction calculation algorithm , here the grey value of a point represents the coner value of the point's normal. Fig. 6 shows the result processed by the fringe curve detection algorithm. It is clear that there are many interrupted gaps in Fig. 6. The final result processed by interrupted gaps identifying and interpolating algorithm is shown in Fig. 7.

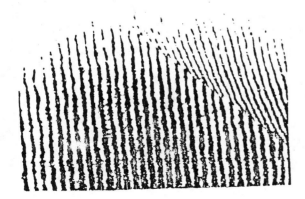

Fig. 4. The original laser interferometric image.

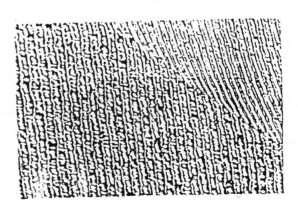

Fig. 5. The result processed by edge detection and inner normal direction calculation algorithm.

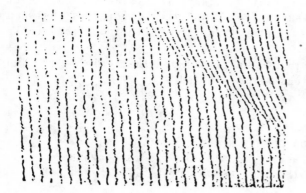 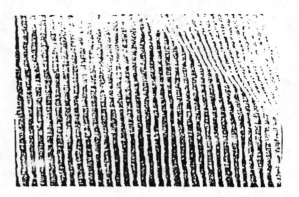

Fig. 6. The result processed by
fringe curve detection algorithm.

Fig. 7. The final result.

5. CONCLUSION

In this paper, we have proposed an effective interferometric fringe curve extration algorithm. The implementation of the algorithm is divided into two steps: first, find out all frnge curve points, then use a curve tracking and interrupted gap linking algorithm to extracted fringe curves. a priori information about the interferometric fringe images is used in the two algorithm. This is the one of main features of our method.

6. REFERENCES

1. He Anzhi et al, Journal of East China Institute of Technology, No.3, 1(1987).
2. D. W. Robinson, Appl. Opt. Vol. 22, No.14, 2169(1983).
3. W. R. J. Funnell, Appl. Opt. Vol.20, No. 18, 3245(1981).
4. W. H. Augustyn, A. H. Rosenfeld, and C. A. Ianoni, SPIE Vol.153, 146(1978).
5. W. Augustyn, SPIE Vol. 192, 128(1979).
6. A. T. Glassman and C. E. Orr, SPIE 81, 64(1979).
7. Y. Seguuchi, Y. Tomita and M. Watanabe, Exp. Mech. 19, 362(1979).
8. J. Y. Yang and K. Liu, Proc. of the IEEE Asian Elec. Conf. 244(1987).

USE OF GRAY SCALE CODING IN LABELING
CLOSED LOOP FRINGE PATTERNS

V. PARTHIBAN and RAJPAL S. SIROHI
Applied Optics Laboratory
Physics Department
Indian Institute of Technology
Madras 600 036, INDIA.

ABSTRACT

Interferograms with closed loop fringe patterns pose numerous difficulties while labeling. Most automatic fringe numbering routines fail while encountered with fringes of complicated shapes. An interactive fringe numbering routine which uses a gray scale coding technique in the form of pseudo coloring to help the machine to identify each order during data input is presented. Each fringe is traced individually and coded with different colors (gray levels) according to its order. The computer is informed of the order of each fringe and the corresponding gray level, so that while it scans the peak contours for data input, it identifies each order even if the fringes are concentric.

1. INTRODUCTION

1.1 Automatic fringe processing

Methods of fringe pattern analysis and interferogram reduction are as important as interferometry, since without analyzing and thereby understanding the data provided by the interferograms the purpose of interferometry is not fulfilled. Application of various electronic aids and computers has made fringe pattern analysis simpler and faster than in the days of manual analysis.

Interferometric data reduction depends on the particular interferometer producing the interferogram, since the method by which optical path difference is achieved varies. But, when interferometric technique are used for repetitive measurements, non-destructive testing or inspection, it is possible to design a comparatively simple fringe analysis system. The repetitive analysis allows for the use of computer. The analysis system can be confined to those required for the measurement in question.

Robinson [1,2] has described a number of specialized fringe analysis procedures for use in engineering application. Robinson[1] developed a method of automatically searching interferograms for the presence of these pre specified faults.

1.2. Interactive fringe analysis

Trolinger[3] has listed out the drawbacks of automatic fringe processing during fringe peak detection. To overcome these drawbacks interactive algorithms for fringe processing have been developed. An interactive algorithm is a semi-automatic algorithm which allows for the operator to direct the computer during processing.

An interactive fringe analyzing system called RIFRAN (Ritan Interactive Fringe Analyzer) was developed by Yatagai and Idesewa[4]. The man-machine interaction is done in RIFRAN with a light pen. When the light pen is pointed at an element on the TV monitor, the computer is interrupted and then reads the x, y coordinates of the light pen hit or the pertinent commands for system control.

Image processing was applied for interactive analysis of fringes by Funnell[5]. The operator interacts with the computer through the key board by directing the machine during the identification of the beginning of the fringe and during the fringe tracing. Interactive fringe analysis has been applied to Moire contourgrams and to test the flatness of very large scale integrated circuit Wafers by Yatagai et al[6,7].

An interactive algorithm should have the following characteristics.

1. Interactive definition of area of interest

2. Editing of results to correct errors in fringe peak detection

3. Interactive fringe numbering

4. Fringe tracing to correct discontinuities

2. INTERACTIVE FRINGE PROCESSING ALGORITHM

2.1 System Configuration

We have developed an interactive fringe processing algorithm using image processing subroutines provided by intellect 100 image processing system. This algorithm has been developed on a PDP 11/23 micro computer.

A schematic of the system configuration is shown in Fig.1. Image acquisition is done by a vidicon. For image display and data Input/Output, Intellect 100 hardware is used. The image processor is based on a frame buffer with 512 x 512 pixels. The image processing system is comprised of the following components.

i) LSI - 11/03 processor
ii) VT103 terminal

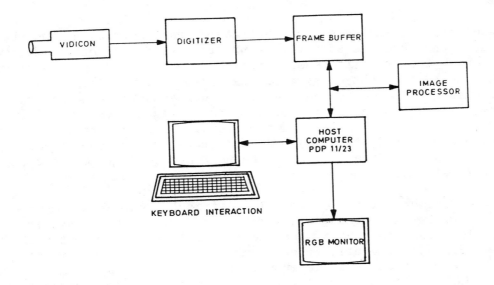

Fig. 1 System configuration

iii)   Intellect 100 frame store
iv)   Image processing package

The frame store is a device capable of storing an entire TV frame in a temporary memory at transmission rate. The incoming analogue signal from the Vidicon is digitized using a fast ADC. The loaded signal is used to represent the luminance of the analogue signal. A complete set of luminance is called a gray scale. A gray scale of 8 bit resolution contains 256 gray levels from black to white. To display the stored image, the digitized frame is passed through an output processor containing a DAC. After the insertion of synchronization signals, the image can then be displayed on a standard monitor.

The computer can be used to manipulate and control the incoming and out going digitized signals, as well as the frame store data. The intellect 100 image processing system is based its frame store, which is capable of storing a complete digitally encoded picture organized as 512 lines by 512 pixels.

The algorithm has six basic components :

a)   Noise reduction and averaging
b)   Thresholding
c)   Peak detection
d)   On-line editing
e)   Fringe tracing and
f)   Labeling

Detailed descriptions of the algorithm have been presented in our earlier papers[8,9]. The use of gray scale coding during labeling is explained here.

2.1 Labeling

For interferogram analysis a set of fringe peak coordinates and the fringe order of each peak are needed. With a prior knowledge of mth order and the direction in which the order increases, the operator can interactively number the fringes. The procedure might consist of the operator identifying a fringe with a light pen or cursor and then typing the fringe number into the computer. In many circumstances, it is possible to overcome the fringe numbering problems by introducing a substantial degree of tilt to the interferogram, so that, the fringes become essentially parallel. In this case, the fringe number increases by unity as we move from one fringe to the next and the fringe numbering can be carried out automatically.

Interferograms with closed loop fringe patterns pose numerous difficulties during numbering[10]. It is possible to have several detected fringe peaks along the same chord with the same fringe number, but with different y-coordinates (Fig.2). Automatic fringe numbering routines usually assign fringe numbers on the assumption of a linear unit increase in fringe number along each scan line. Such a routine will fail when a closed loop fringe pattern is encountered during evaluation. Ruixiang Liu and Birch[94] have proposed a method to overcome this problem. According to this procedure, the operator must enter for each scan line I, the starting values of four parameters. These are the initial fringe number H(I), the number of loops in the scan line B(I), the maximum fringe number K(I) and the number of detected peaks at the maximum fringe number A(I). With this information, the computer then proceeds to number the detected fringes automatically. This process is repeated for each chord line.

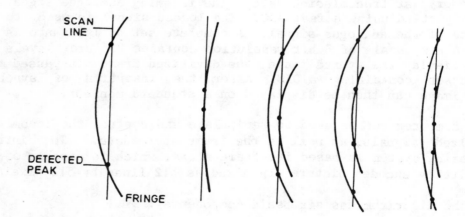

Fig. 2   Examples of closed loop fringe patterns showing the scan line
cutting the same order more than once.

A much simpler procedure has been developed in the interactive algorithm to number closed loop fringe patterns. To help the machine to identify the fringe order, the operator pseudo colour codes each contour line by changing the gray value of each fringe contour by tracing them individually and tells the machine the gray value and the order it represents.

2.2 Fringe tracing

The Fringe tracing algorithm is also highly interactive and requires operator's guidance at all time. In fringe tracing, the cursor is placed at the start of a fringe and as the operator tells the computer to proceed, it scans the five pixels surrounding it two adjacent to it on the same line and three immediately below it on the next line. The machine then follows the fringe and when there is a discontinuity or ambiguity, it pauses and asks for operator's guidance. The operator then has to decide on the correct path and guide the machine to proceed. As each order is being traced, the operator tells the machine the order number and the computer assigns a gray scale corresponding to the order and then changes the gray scale of the particular fringe contour. Thus the computer can identify each fringe order by its gray value and store it in a data file. This technique will be successful when complicated fringe patterns are processed. When the fringes are in the form of loops, each order can be distinguished by the computer by its gray value.

Fig.3(a) shows an interferogram with closed loop fringes and Fig.3(b) shows the extraction of peak contours.

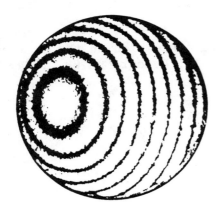

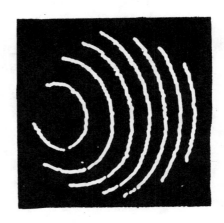

Fig. 3(a) Closed loop fringe pattern

Fig. 3(b) Peak contours of the closed loop fringe pattern

3. REFERENCES

1. D.W. Robinson, 'Role of automatic fringe analysis in optical metrology,' Proc. SPIE, **376**, 20 - 5 (1983).

2. D.W. Robinson, 'Automatic fringe analysis with a computer image processing system', Appl.Opt., **22**, 2169 - 76 (1983).

3. D. Trolinger, 'Automated data reduction in holographic interferometry', Opt.Eng., **24**, 840 - 842 (1985).

4. T. Yatagai, S. Nakadate, M. Idesawa and H. Saito, 'Automatic fringe analysis using digital image processing techniques', Opt.Eng., **21**, 423 – 5 (1982).

5. W.R.J. Funnell, 'Image processing applied to the interactive analysis of interferometric fringes', Appl.Opt., **20**, 3245 – 49 (1981).

6. T. Yatagai and M. Idesewa, 'Automatic fringe analysis for moire topography', Opt. and Lasers in Eng., **3**, 73 – 83 (1982).

7. T. Yatagai, S. Inber, H. Ndakano and M. Suzuki, 'Automatic flatness tester for very large scale integrated circuit wafers', Opt. Eng., **23**, 401 – 5 (1984).

8. V. Parthiban and R. S. Sirohi, 'Interactive fringe processing using pseudo coloring', Presented at "FRINGE 89" the first international workshop on automated processing of fringe patterns, Berlin, G.D.R., April 25 – 28 1989.

9. V. Parthiban and R.S. Sirohi, 'Interactive fringe processing algorithm for interferogram analysis', Accepted for publication in the Optics and Lasers in Eng. (U.K.).

10. Ruixiang Liu, K.G.Birch, 'Experiments and discussion on the analysis of interferograms with closed-loop fringes', NPL Report MOM **68**, 1984.

An Assessment of Some Image Enhancement Routines for use with
an Automatic Fringe Tracking Programme

J.C. Hunter, M.W. Collins, & B.A. Tozer

The City University, London, England.

ABSTRACT

An automatic fringe tracking programme has been developed for use with holographic interferograms obtained for heat transfer and compressible flow studies at the City University Thermofluids Laboratory. This paper reviews the image enhancement routines which were considered for use to prepare images prior to the application of fringe tracking and fringe ordering procedures. Their relative performances are assessed qualitatively and the conclusions which were reached in these studies are presented. Data pre-processing aspects such as image masking and the selection and implementation of common point or fiducial marking are also described.

An important aspect of the work described has been the necessity of operating with an optimised level of image enhancement, trading off processing time and hardware costs against usable image quality. Maximum entropy and statistical enhancement methods were rejected in favour of more direct and speedier methods.

The heat transfer experiments were carried out with more primitive equipment than that used for the compressible flow studies, yielding poorer quality raw data, so that a much more extensive image enhancement regime was required before adequate data was available for the fringe detection routines. The paper details the different routines required in these two cases, illustrating how experimental sophistication can be traded against image processor costs and processing time.

1. INTRODUCTION

It has been recognised for many years that optical interferometry has much to offer to engineers in the various branches of mechanical engineering. The study of structural distortions and vibrations and problems involving fluid flow can benefit enormously from the ability of interferometric techniques to provide detailed whole field data on these phenomena with almost unparallelled sensitivity to surface movement or change in fluid density.

Despite the simplification in the technology of interferometry resulting from the introduction of coherent laser sources and the invention of holography, the engineering community has been slow to adopt interferometry for routine shop floor applications. One of the reasons for this lukewarm response is undoubtedly the fact that the results obtained must usually be analysed by hand and, generally, are only interpreted by highly skilled research engineers. Even so, results are usually only interpreted qualitatively and no routine methods are available to produce whole field quantitative data from interferometry. As a consequence the study reported here is only one of a number published recently which are aimed at enabling the engineer to obtain whole field quantitative data from interferograms, quickly and painlessly. This paper deals with the treatment of raw data prior to the application of a fringe tracking routine and shows how experimental sophistication can be traded against image processor costs and processing time.

Before a fringe system can be analysed automatically, or semi-automatically, two stages of data pre-processing are usually needed. These are, respectively, data preparation, in which the fringe field data is formatted to allow the use of fringe analysis routines, and image enhancement which is usually found to be necessary to minimise the loss of image quality which normally results when the raw data is accessed and transferred to a digital data file for computer analysis.

2. DATA PREPARATION

2.1. General

The data preparation stage includes such procedures as masking off physical objects appearing in the fringe field; defining the fringe field if necessary; and defining common points for location and scaling purposes if for any reason a magnified portion of the fringe field has to be analysed separately and then merged into the fringe data.

2.2. Masking Techniques

In most fringe field analysis situations it is necessary to define the extremities of the fringe field so that the fringe analysis routines may recognise the boundaries within which they must operate. It may also be necessary to mask off physical objects obtruding on the field, so that the fringe analysis routines may ignore them.

Three approaches to the masking problem were considered:

1) Segmentation: The regions to be masked are detected by their grey level. This method is simple, but cannot be applied unless the regions to be masked have a unique grey level range not shared by any portion of the field to be analysed. This is unlikely to be the case where interferometric fringes are present.

2) Boundary Tracing: Suitable algorithms have been developed (see e.g. Pratt [1], Castleman [2]), which work on maximal gradient principles. However they are very sensitive to noise, and even a small amount of noise can send the tracking off the boundary. It was concluded that these techniques are useful only when there is a low signal-to-noise ratio, or where some form of operator interaction is provided to correct possible derailment.

3) Template Matching: In this technique a replica of the object to be masked is compared with all the objects in the image field. This method is inapplicable because it requires segmentation of the whole image field. Furthermore, in the City University work, the objects in the image field did not have a reproducible shape from record to record so that the template would not necessarily match the area to be masked.

The masking routine developed in this work is illustrated in the algorithm of Fig.1, and operated as follows:

a) Re-assign all grey levels, leaving a unique value (level of 100 assuming a file with 255 grey levels) available for allocation to masked areas. All routines ignore this unique level and image enhancement routines do not allocate to this level.

b) Implement an edge routine to mask off straight line boundaries. An operator controlled cursor defines the start and finish of each section. The direction to mask off is specified by the operator.

c) Mask off more complicated shapes. The basic geometry of shape is written into file with a list of basic geometric descriptors. The operator controlled cursor is used to locate the necessary geometric descriptor points, for example, in Fig.2, the basic geometry is a rectangle, with a semi-circle at each end. Geometric descriptors are length, width and angular orientation of the blade. The cursor defined points on the image data for the first blade were: 2 on upper surface (describes angular orientation), one on the lower surface (gives thickness) and one point at each end to describe the blade position. Subsequent blades needed only one cursor point on the blade end.

2.3. Common Point Description

It is still usual to carry out image processing and fringe tracking routines with processors using only 512 x 512 pixels, or fewer, with a grey scale of 8 bits or less. These figures are inadequate for the analysis of anything but relatively simple fringe fields, except by a process which involves the acquisition and processing of data in a number of segments, and their re-merging after analysis. For these cases it is necessary to have a procedure for common point description so that the merging of processed data can be achieved with precision.

Two different techniques have been used in the work reported here, to meet differing requirements. In one case, Fig.3, a kinematic plateholder was used to provide an overlay of fiducial marks on the holographic image. Using these, a 512 x 512 pixel by 8 bit frame grabber and a 1024 x 1024 x 8 bit data file it was possible to process, track and number the fringes over the entire scene, and to view the output on a 512 x 512 pixel by 8 bit image using a windowing routine which allowed the operator to select a window covering any part of the image file.

In the other case, which involved a study of heat transfer from a heated undulating surface (Fig.4) an increasing density gradient meant that portions near the heated surface had to be magnified in order to provide sufficient resolution. In this case common points were defined within the image itself, using easily recognised features. Once again the processed data could be merged within a single data file.

3. IMAGE ENHANCEMENT

3.1. General

After masking the image it is necessary to perform some image enhancement before undertaking fringe tracking because the image may have already been substantially degraded during the data acquisition and digitisation stages. There is no general unifying theory of image enhancement at present because there is no generally accepted standard of image quality to serve as a criterion for the design of a processor. In practical situations it is necessary to experiment extensively to find the most effective technique, and this must often be a "tuneable" method whose parameters may vary from place to place in the image.

This work was directed to two very different fluid flow regimes. One (Fig.4), as already noted, was a heat transfer experiment in which relatively few fringes occur at ever decreasing intervals as the heated surface is approached. The results obtained in this work lent themselves to a fringe analysis system operating in one dimension only (perpendicular to the surface). The second regime (Fig.5) was a study of compressible fluid flow in an incidence cascade rig designed for the investigation of off-design effects on low pressure final stage tip section blading. Clearly these two applications presented very different problems for automatic analysis, some of which have already been touched upon in Section 2 above.

The holograms obtained from the heat transfer experiment were of a much lower quality than those obtained later on the cascade rig, and the digitisation equipment available for the former experiment was also less sophisticated and provided less control over the digitisation step. In consequence, although the same menu elements were available for image enhancement routines in both cases, the heat transfer results needed a much more comprehensive enhancement. The enhancement menus found to be most useful, in each case, are summarised in Table 1. All the enhancement routines are written so that they will ignore the masked regions and will not allow the unique value of grey scale defining the masked regions of the fringe field to appear in the fringe field itself.

Basically the image enhancement routines are aimed at achieving improvements in four important respects. These are:

a) Maximising contrast by using the available grey scales to fill the dynamic range.
b) Reducing stochastic noise in the image.
c) Enhancing edges to improve fringe visibility.
d) Equalising background intensity to correct for spatial variations in detector response and the approximately Gaussian beam intensity profile of the laser source.

Whilst the correction of each of these faults presents its own special difficulties, the ability to correct (b) and (c) above is limited by the fact that each requires processes that exacerbate the problems caused by the other.

3.2. Grey Scale Enhancement

Three methods of grey scale enhancement were employed. Two of these (i) "linear histogram stretch" in which the distribution of grey scales is modified to give a linear spread of values between white and black and (ii) "histogram equalisation" which employs the operator:

$$F(i) = Min(255, [\Sigma_{j \leqslant i} H(j) \times 256/N]) \tag{1}$$

to distribute the histogram, are conventional. In addition the novel technique of (iii) "histogram split" was employed. In this technique, which is especially suited to a fringe field, a grey level distribution which is observed to be bimodular, or double peaked, in form is split in the middle and the two halves are allocated to either end of the grey scale (Fig.6).

Grey scale enhancement by histogram split can be especially beneficial when dealing with fringe fields, but it is necessary to exercise some caution when using it because artificial contouring may be introduced by such sudden grey level discontinuities. For this reason the image data should be checked by taking intensity distribution plots across the fringe field to ensure that the magnitude of any artificial contouring is small compared to the sinusoidal variations of the fringes. It should be noted that the higher quality fringes and better data acquisition equipment used in the compressible flow studies could be handled satisfactorily by the linear histogram stretch method, which is much less likely to introduce artefacts into the fringe data.

3.3. Noise

Noise occurs in the fringe field for a number of reasons varying from electrical noise to speckle and processing noise. We deal mainly with stochastic noise here. Since stochastic noise appears as a high frequency phenomenon, whilst fringes usually have low to medium spatial frequency, it follows that low pass filtering should produce a significant improvement. However the elimination of high frequencies has an undesirable effect on edge sharpness.

A median filtering technique developed by Tukey [3] was employed in which the intensity at a point is replaced by the median intensity value in a designated surrounding neighbourhood. In one dimension the process removes isolated spikes without distorting edges or gradual intensity variations. In two dimensions its effect is very dependent on the shape of the neighbourhood filter. A cross neighbourhood (Fig.7) which was used for the heat transfer work, leaves horizontal and vertical lines but destroys diagonal lines. The technique was found to be as effective as low pass filtering for noise removal, whilst maintaining better fringe contrast.

3.4. Edge Enhancement

High pass spatial filters or non-linear spatial filters (e.g. Sobel operators) can be used. Alternatively, a linear filter working on the discrete differentiation principle can be used. However edge enhancement is of limited use in fringe analysis systems because of the increased noise levels which result from the enhancement of high frequencies.

3.5. Background Intensity Equalisation

Because the background function is modulated by the fringe function it is not possible to model the laser beam's Gaussian distribution for correction purposes. A method similar to that of Becker and Yu [4] was used here for the heat transfer analysis work. The digitised image was divided up into a number of small square regions, typically 1024. The average grey level was determined over each small square (excluding contributions from masked regions) and these values used to form a coarse grid of node points. An average (grey level intensity) surface was then fitted over this grid using the bilinear interpolation technique shown in Fig.8, and the surface then used to provide a relative correction factor for each point. A corrected image is shown in Fig.9.   In the later compressible flow study the background equalisation technique was incorporated into the binarisation stage of the fringe tracking routine, which allowed a more efficient algorithm to be developed. This will be described elsewhere (Hunter et al. [5]).

4. CONCLUSIONS

Techniques for the preparation and image enhancement of interferometer fringe fields, prior to the application of fringe tracking routines, have been described. When poor quality data is used there is a requirement for a greater degree of image processing, requiring not only a longer processing time, but also a far higher degree of operator judgement, and a higher risk that results will contain misleading artefacts.

An algorithm has been given describing a masking procedure which operates with only minimal operator input.

Novel methods of contrast enhancement and background intensity equalisation have been described.

5. ACKNOWLEDGEMENTS

This work was originally sponsored by a U.K. Science and Engineering Research Council/Central Electricity Generating Board CASE Studentship EB 052.

6. REFERENCES

1. W.K. Pratt. _Digital Image Processing_, Wiley-Interscience, 1978.

2. K.R. Castleman. _Digital Image Processing,_ Prentice-Hall Inc., 1979.

3. J.W. Tukey. _Exploratory Data Analysis,_ Addison-Wesley, Reading, Mass., U.S.A. 1971.

4. F. Becker & Y. Yu. "Digital Fringe Reduction Techniques Applied to the Measurement of Three-Dimensional Transonic Flow Fields". Opt.Eng., 24, p.249, 1985.

5. J.C. Hunter, M.W. Collins, & B.A. Tozer. "Scheme for the Analysis of Infinite Fringe Systems". Proc. SPIE Conf. 1163, San Diego, U.S.A. 1989.

Table 1

The Image Enhancement Menus Used in this Work

Compressible flow system	Heat transfer system
(i) Contrast enhancement by linear histogram stretch [G]	(i) Low-pass filtering (smoothing) [G/L]
(ii) Low-pass filtering (smoothing)	(ii) High-pass filtering (edge crispening) [G/L]
(iii) High-pass filtering (edge crispening [G/L]	(iii) Histogram stretch |contrast (iv) Histogram equalise |enhance- 　　　　　　　　　　　　　| ment (v)　Histogram split 　| [G]
	(vi)　Intensity equalise [G]
	(vii) Median filter, cross neighbourhood [G/L]
	(viii) Laplacian edge-detection [G]
	(ix)　Sobel edge-detection [G]

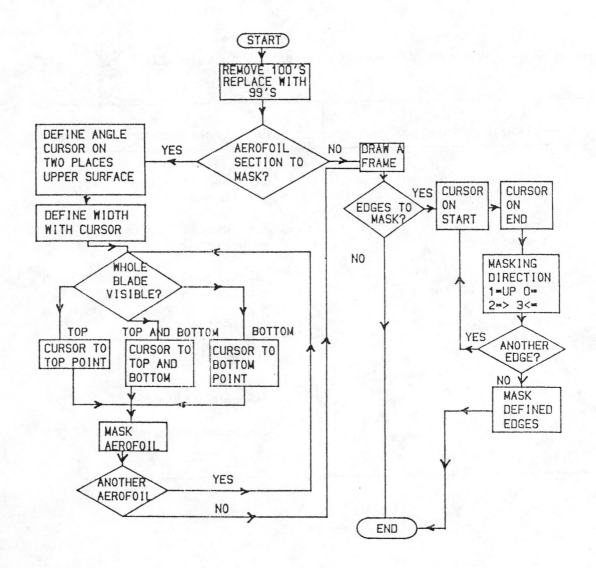

Figure 1

Flow Diagram of Masking Procedure

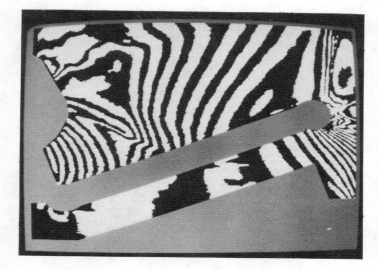

Figure 2

Example of masked Image.
Note unique masking grey level = 100

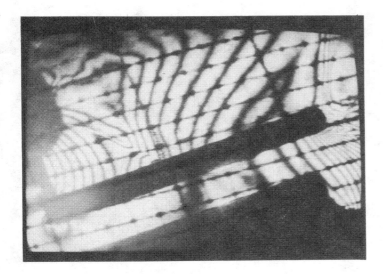

Figure 3

Fiducial Marking System
Used for Common Point Description

Figure 4

Heat Transfer Interferogram.
Common Points can be defined by Features

Figure 5

Compressible Flow Interferograms
Must be Segmented for Digitising
with 512 x 512 pixels

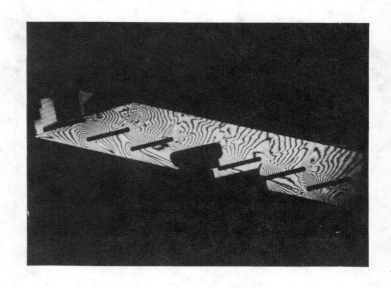

Figure 6

Contrast Enhancement by Histogram Split

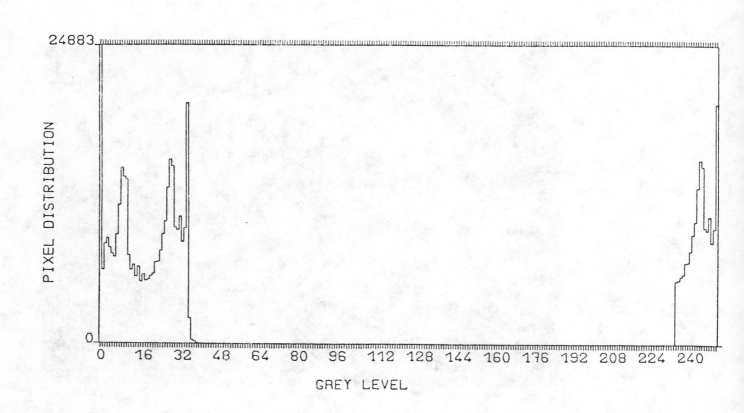

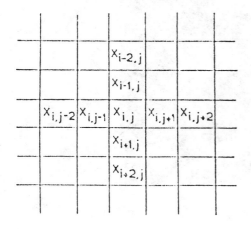

$Y_{i,j}$ = median of $\{\,X_{i,j-2},X_{i,j-1},X_{i,j},X_{i,j+1},$
$$X_{i,j+2},X_{i-2,j},X_{i-1,j},X_{i+1,j},$$
$$X_{i+2,j}\,\}$$

Formation of pixel in location i,j of filtered image

Cross neighbourhood of Xi,j in the original image

Figure 7

A Median Filter,
Cross Neighbourhood

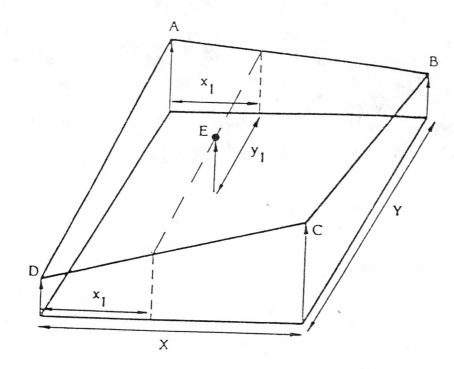

Figure 8

Background Intensity
Equalisation by bi-Linear
Interpolation
E =
$A(1-x_1/X -y_1/Y + x_1/X.y_1/Y)$
$+ Bx_1/X(1-y_1/Y)$
$+ Cx_1/X.y_1/Y + Dy_1/Y(1-x_1/X)$

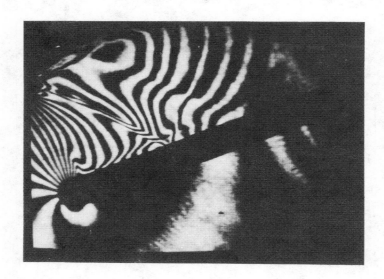

Figure 9

Top. Raw Image
Bottom. Image after
Equalisation and
Histogram Split

A QUASI HETERODYNE HOLOGRAPHIC TECHNIQUE AND AUTOMATIC ALGORITHMS FOR PHASE UNWRAPPING

D.P. Towers,

T.R. Judge,

P.J. Bryanston-Cross

Department of Engineering, Warwick University, England.

ABSTRACT

This paper describes recent developments on methods for quantitative analysis of holographic interferograms. Two areas are considered : the holographic system and the automatic processing of fringe images.

The holographic system is based on a quasi heterodyne technique. Methods for reconstructing the image with a single beam are presented. This produces a continuously variable phase difference between the two images of the object. An analysis of the errors has been performed for the reconstructed phase maps.

A tiling approach and graph theory have been applied to the fringe field processing problem. This has produced an automatic method for analysing fringes forming an optimum solution for a phase or height map.

1.0 Introduction

Data reduction techniques for holographic interferograms are of increasing importance for a variety of industrial measurements. Typical applications are in vibration analysis and holographic contouring. The development of the heterodyne and then the quasi-heterodyne techniques have provided a means for making phase measurements at any point in a fringe field (ref. 1). In particular the quasi heterodyne method allows this to be done in a microcomputer via a CCD camera and digital framestore (ref. 2). Hence rapid, on site analysis of holographic fringe images has become possible.

For industrial holography to provide an efficient tool, the fringe patterns should be analysed automatically with minimum human intervention. In some specific cases algorithms have been demonstrated for automatic fringe analysis (ref. 3). However in these examples either prior information on the type of fringes formed by the system is utilised, or identification of a particular fringe feature is required. A general fringe pattern requires an algorithm to overcome the problems of random noise and object discontinuities. The approach formed to solve this problem splits down the image into a set of tiles (ref. 4). On an individual basis the heights for the tiles are solved incorporating tests for continuity across the area of the tile. A tree is then formed linking all the tiles to re-constitute the phase or height map for the whole image.

Experiments have been performed for both holographic and ESPI fringes. The holographic images are produced by a dual reference beam method where a spatially varying phase difference is formed between the two images. On reconstruction of the hologram with a single beam, the phase in the image can be varied by a change in viewing point. Phase calibration is achieved from the reconstructed image. The ESPI images are produced by phase stepping in the reference beam with a fibre optic wrapped around a PZT (ref. 5). These images were obtained from the Rover Group research centre at

Gaydon. Methods for image preparation are presented for both holographic and ESPI fringes.

For both types of fringe images, three phase views are captured. These are combined mathematically to produce fringes with a tangent function of phase across them (ref. 6). Therefore the slope of the surface can be determined in the software analysis.

2.0 Dual Reference Beam Holographic Method

2.1 Description of the Approach

A schematic diagram of a dual reference beam holographic arrangement is shown in Fig. 1. The method developed uses two reference beam sources close together in space. This reduces distortion affects due to inaccurate hologram repositioning (ref. 1,7,8). Errors due to different recording and reconstruction wavelengths are also minimised by this method. The holographic image is reconstructed with a single beam.

From an analysis by Vest (ref. 9), the phase difference, δ, measured by an observer at a point Q in the hologram plane is

$$\delta = (k_2 - k_1).L$$

if the same reference beam is used for both exposures. However if two distinct reference beams are used, R1 and R2, with phases of ϕ_{R1} and ϕ_{R2} at Q respectively then the measured phase difference becomes

$$\delta = (k_2 - k_1).L + (\phi_{R2} - \phi_{R1}).$$

The fringes in the reconstruction now represent the vibrational mode modulated by the phase difference due to the two reference beams.

If R1 and R2 are plane waves the phase difference $\phi_{R2} - \phi_{R1}$ will vary linearly across the hologram (i.e. Michelson fringes are formed). With spherical beams the phase difference will also vary across the hologram , e.g. from Q to Q', due to the formation of a Young's fringe pattern. In either case, the fringes can be scanned over the object as the reconstructed image is viewed through different parts of the hologram. Therefore a CCD camera mounted on a linear traverse can examine the fringes at any phase angle.

2.2 Holographic Arrangement

The dual reference principle has been demonstrated using a plane sheet caused to vibrate with an impinging air jet. Holograms were produced using a double pulsed ruby laser. The optical arrangement is shown in Fig. 2. A collimated sample beam illuminates the vibrating sheet. Back scattered light is combined with the reference beams at the holographic plate to produce a transmission master hologram.

An optical switch provides two different reference beams to form a double pulsed hologram (ref. 4). Ten percent of the laser beam is used for the reference and is in turn split into two equal halves. Each half proceeds to a Pockels cell and polarising cube beamsplitter. The voltage applied to the Pockels cell determines the direction of polarisation of the beam transmitted through it. Hence a switching of the beam path is produced at the polarising cube beamsplitter. The two vertically polarised beams which can be produced from the system are re-combined at a second 50/50 beamsplitter. From the optical switch the reference beams follow a nearly common path through a set of path length matching mirrors. A small positive lens expands the beams over the holographic plate. The beam separation was set to approximately 0.1 mm measured at the focus of the beams produced by the lens.

The switching efficiency of the system is mainly dependent on Pockels cell alignment. This can be measured from the intensity of the vertical and horizontal components of the beam which are separated by the polarising cube beamsplitter. An efficiency of better than 99% has been achieved.

Electronic hardware was developed to control the firing of the pulse laser and external Pockels cells. This allows the system to be used down to a pulse separation of 0.5 microseconds.

Interferograms were produced using a pulse separation of 250 microseconds and developed in Neofin Blau.

2.3 Image Reconstruction

The hologram is reconstructed using a single beam, see Fig. 3. A CCD camera mounted on a linear traverse views the virtual image. The CCD output is received by a digital framestore card and a monitor. The framestore has a resolution of 512*512*8 bits giving 256 grey levels. As the camera traverses the hologram, the fringes are swept across the object. The framestore also detects a vertical shift in image position. The traverse can be calibrated by comparing two fringe images of 360 degrees phase difference. The vertical translation of the image with traverse distance also needs quantifying to allow two different phase views of the object to be overlaid in the framestore.

Traverse calibration is achieved by defining a feature in a reference image. This is compared to every possible position the feature can take in a new image using a chi squared test. As the image only translates vertically the feature must occur between fixed horizontal limits. Therefore each position of the feature is considered in a vertical strip and chi squared values computed. The minimum chi squared value yields the pixel shift between the two images. This process is repeated for different traverse positions. A minimum chi squared value with respect to traverse distance results in the calibration with respect to phase.

An image at any required phase can now be found by linear interpolation of the traverse distance for 360 degrees phase shift and overlaid over the reference.

2.4 Sources of Error in the System

The dual reference beam method described above has implicit errors due to reconstruction of the holographic image with a single beam. The errors can be estimated using the first order approximation developed by Hariharan (ref. 10), see Fig. 4. From this analysis the z co-ordinate of a reconstructed point is given by

$$Z_3 = \frac{Z_O . Z_R . Z_P}{Z_O . Z_R + \mu . Z_P . Z_R - \mu . Z_P . Z_O}.$$

The subscripts o, r, p represent the object, reference and reconstruction sources respectively; μ is the reconstruction to recording wavelength ratio.

A measure of the phase errors in reconstructing a double pulsed hologram can be obtained by considering the Z co-ordinate of points on the two surfaces. Any distortions in the X and Y directions will be neglected. To produce no errors, μ must be equal to 1.0 and the two reconstruction beams must be precisely the same as the original reference beams. Also, if one surface is translated in depth relative to the other, with no further distortions, the variations in phase across the fringe pattern will remain unaltered (providing decorrelation does not occur).

The equation above implies that the reconstructed Z co-ordinate is a non-linear function of the object Z co-ordinate. Simulations have been performed to examine the error variations using an inclined flat plate which is considered to have a pure translation, ΔZ, between the two pulses of 10 microns (μm). In the ideal case the reconstructed image will have no variations in phase across the plate as a constant ΔZ has been applied to all points. For the single reconstruction beam method described above, sample coordinates used for simulation are $Z_{R1} = -1500$, $Z_{R2} = -1500.01$, $Z_P = -1500$, and a range of object depths from -2000 to -2100 (dimensions in mm, see Fig. 4). For $\mu = 1.0$ the actual distance produced between the two surfaces is 27.77 μm with a variation of 1.82 μm over the range of depths examined. This can be considered as an increasing offset, or ramp, with added non-linearities. If the ramp is assumed to be linear it can be removed from the phase map formed. This produces the same distance between the two surfaces and a reduced variation of 11.1 nanometers over the depth range of 100 mm. For the holograms produced the wavelength ratio of the Helium-Neon (reconstruction λ), to the ruby (recording λ) is 0.9121. In this case the errors are an offset of 23.91 μm and a variation in phase over the depth of the object of 5.68 nanometers, after ramp correction. Therefore the wavelength ratio compensates for some of the errors in incorrect reconstruction. This represents a phase error of less than 1 part in 100. In the general case the reconstruction depth will not be identical to that of either of the reference beams. Simulations with the following figures have been performed : $Z_{R1} = -1500$, $Z_{R2} = -1500.01$, $Z_P = -1600$, $\mu = 0.9121$ with the same range of object depths. This produces a phase variation of 1.75 μm which reduces to 10 nanometers, or 1 part in 70 phase error, after ramp correction.

A second major source of error is incurred when the fringe images are shifted in the framestore. This is currently performed to an accuracy of 1 pixel. The worst case error will occur for the fringe covering the least number of pixels from peak to peak in the direction of the translation. For the particular hologram shown (see Fig. 11) the minimum number of pixels covered by a fringe is 20. Hence a phase error of 1 part in 20 is produced.

There are four other sources of errors which are listed below :

i) Linear interpolation has been assumed between the two traverse calibration points, which are 360 degrees apart. This produces a phase error of 1 part in 3000 for the geometry in Fig. 4.

ii) An experimental error due to the repeatability of the minimum chi squared measurements gives a phase error of 1 part in 200.

iii) The intensity in the cosinusoidal fringe images is quantised to 8 bits. This causes a maximum error of 1 part in 1200 for the phase measurement. The calculation is based on the arc tan function of intensities given in the following section.

iv) In the dual reference beam case, the two beams add, forming Young's fringes which are projected onto the holographic plate. This leads to uneven diffraction efficiency. Hence non-uniform image intensity can result. In this case the variations are small and have not been quantified.

An estimate of the phase error that could be produced from this dual reference beam system is 1 part in 19, assuming that ramp correction has been applied. This corresponds to a measurement accuracy of 20 nanometers. Currently the limiting factor is in the pixel shift applied to the fringe images in the framestore. This error can be reduced by incorporating a sub-pixel shift algorithm.

3.0 Image Preparation for Software Analysis

The quality of a fringe image can be assessed by inspection of the fast fourier transform (FFT) of a scan line. The high frequency components represent noise, mainly due to laser speckle, whilst the lower frequencies represent the desired information about the fringe separations. Both the holographic and ESPI images were filtered before being used in the analysis software. An averaging filter was applied where the value at a pixel is replaced by the mean of the values of its eight nearest neighbours. This has the affect of smoothing out the high frequency noise in the image. With successive applications of the averaging filter, the value at a point becomes influenced by pixels further away hence increasing the smoothing affect.

The optimum number of filter processes applied was experimentally determined by examining the signal to noise ratio for a row of the image. A measure of the signal to noise ratio was

calculated by evaluating the median for the FFT. Any value greater than the median was taken to be signal, and any value less than or equal to the median taken as noise. When a maximum value of signal to noise ratio was reached the optimum number of passes was found. For the holographic images a single pass of the averaging filter was required, whereas four passes were needed for the ESPI fringes. The FFT's are shown in figs. 5a and 5b for the holographic and ESPI fringes respectively (with the D.C. term removed from the plot). The top curve in each graph represents the smoothed data and the bottom curve the raw data.

After averaging, each set of fringe images are normalised before entering the software analysis. The first operation performed is to create the 'wrapped' phase map. This is calculated using the expression (ref. 2)

$$\phi = \tan^{-1}\left[\frac{\left[I_{2\theta}-I_0\right].\left[\cos\theta-1\right]-\left[I_\theta-I_0\right].\left[\cos2\theta-1\right]}{\sin\theta.\left[I_{2\theta}-I_0\right]-\sin2\theta.\left[I_\theta-I_0\right]}\right]$$

Where θ is the phase step between the images (120 degrees and 90 degrees for the holographic and ESPI images respectively), I_0, I_θ, $I_{2\theta}$ are the pixel intensities in the three images, and ϕ is the interference phase at the pixel. The 'tan' fringes show a graduation in intensity from black at one edge to white at the other. This represents a phase change of 180 degrees. One fringe width indicates a fixed displacement related to the wavelength of light used. The phase allows a fraction of that displacement at the pixel point to be calculated. Therefore it is possible to measure height using pixel intensities. However the measurements are made with respect to the start of each fringe. To calculate the absolute height the tan fringe image must be 'unwrapped'.

4.0 Phase Unwrapping

4.1 The Processing Aim

The aim of processing is to automatically produce a height map of the surface described by the tan fringe field. The processed output should contain a displacement for each pixel in the frame relative to some single fixed origin.

4.2 The Processing Problem

In order to produce an automatic processing system, techniques must be developed to cope with the problems posed by the general fringe field.

The general fringe field may contain,

i) Noise, which is introduced when the input images are recorded.

ii) Points of low modulation, which contain insufficient phase information.

iii) Discontinuities, resulting from gaps in the fringe field.

iv) Non linearities in intensity, which may be brought about in the holographic reconstruction process.

v) Ambiguous areas, resulting from an insufficient pixel resolution to 'see' all of the fringes.

All of the above may occur at any point in the field and interfere with construction of the height map, which requires a continuous path across the field.

4.3 Adoption Of The Tiling Method

It is difficult to design an algorithm to construct the height map when one considers the field as an indivisible unit. However, noting that most distorting affects are localised suggests that processing on a regional basis might help. If regions are processed separately a distorted one cannot directly effect the entire map. The approach adopted, therefore, relies upon division of the complete field into smaller field areas, or tiles which may be processed individually.

5.0 Overview Of Tile Processing

5.1 Consistency In The Field

The processing of individual tiles relies on consistency in the field over small areas, that is consistency in the fringes that cross each tile. If a tile has sufficient, then it may be solved.

5.2 Connecting The Map

Once each tile has been processed a connection tree is found that will form the map with optimum confidence. In order to do this graph theory is employed to construct a minimum weight spanning tree (section 3).

5.3 Tiles That May Not Be Processed

If the consistency of fringes across a tile is low then processing of the tile will fail. This is desirable as it removes bad data from the fringe field. Nothing is lost, provided enough tiles remain to construct the spanning tree.

5.4 Excluding Tiles From Processing

Tiles with a high percentage of low modulation points are immediately excluded from processing. This considerably eases the task of automation, as for example static regions of the field are rejected.

The operator may exert control over the region of the field to be analysed by defining a polygonal boundary, with respect to one of the input images. An example of this is shown in Fig 6 along with an illustration of the way in which the tiles cover the field.

6.0 Connecting The Tiles Using A Minimum Weight Spanning Tree

The strategy for connecting the tiles to form the map will now be examined. This is independent of the methods that might be utilised in the solution of the tiles. The solution used here has been described in ref. 4. The connection process has been updated since this paper and now has a more formal basis.

The strategy employed is based in graph theory. This permits the map to be solved with the greatest confidence. The problem of processing the field is one that may only be solved as far as the data available will permit, almost inevitably the field will contain defects. The spanning tree attempts to minimise the affect of these defects. It does this by attaching the tiles containing them, to the map, only when there are no other tiles of better quality. That is by connecting areas of the field in a priority scheme such that the areas, that are more likely to be solved correctly, are combined first.

At entry to the arrangement stage of processing, there exists a set of tiles containing height information. Each tile holds the solution of the wrapped phase map (or tan fringe), for the part of the fringe field over which it has dominion. The solution contains measurements of height for the tile with respect to an origin point inside it. The problem is to adjust these heights, by adding appropriate offsets, so that they are measured with respect to a single origin for the entire fringe field. Fig 7 shows how this is achieved by following the branches of a tree.

The tiles are positioned in height so that they follow a spanning tree. The change in height across a tile, from one edge to the other, as the tree is traversed from the root tile sum to give the height offset of tiles within the tree relative to the root. Defective tiles are forced to the tips of the tree branches so that they do not distort the tree.

The problem is analogous to the solution of a jigsaw puzzle. A jigsaw puzzle starts with a single piece (the root). Pieces are then successively added around it until the whole picture is made. Each tile may be thought of as a jigsaw puzzle piece that must be fitted to form the completed fringe field. At each stage the piece that is thought to fit best is added next.

The edges of the tiles have profiles that must be matched up. However, contrary to the case with the jigsaw there is no guarantee that they match exactly. A confidence level must be found that indicates how well each tile fits.

This confidence level is determined by a software routine that performs an inspection of the edge profiles of adjacent tiles, using a four pixel overlap. The joining order is partly determined by this confidence level and partly by fringe density, as will be seen in section 6.2.

6.1 The Minimum Spanning Tree And Faulty Tiles

Suppose a tile in the tree is at a point in the field where the fringes are very fine. It may be that a fringe is so fine that it passes undetected by the digitising system. The fringe may then grow wide enough to be visible. The result is an error within the tile. Members of the tile tree that follow the missing fringe will be in error by a fringe width, that is out of sync with the other tiles.

Fine fringes occur in parts of the field where the fringe density is great. It is more likely that a fringe becomes too fine for the pixel resolution of the digitiser in one of these areas than in an area of low density fringes. It is, therefore, more likely that an erroneous tile will be found in an area of high density fringes than in one of low density.

It is desirable to connect tiles in low density areas first in order to minimise problems of this nature. The priority ordering is extended to compose both a factor for fringe density and tile boundary confidence. The data on the density of fringes in each tile is available as a by product of the tile solutions.

6.2 Accounting For Fringe Density

The fringe density factor is a value for the fringe density along the tile boundary. This may be calculated from the average of the fringe densities of the tiles on either side of the boundary.

A continuous priority scale is required. The best way to equally bias two factors is to form their product. However, the confidence level and the fringe density are opposite in sense. That is, the confidence level increases with a good route (0 -> 1) and the fringe density decreases. This is simply adjusted by forming the product from (1 - confidence) and the fringe density. This means a low value for the product indicates a good route and a high value a bad route.

One final amendment. If no fringes cross a tile, that is the fringe density is zero, then the product and hence the weightings become zero, which gives no priority information. In order to avoid this the fringe densities are incremented by one before forming the product. If w is the priority or weighting factor, c is the edge confidence, and the fringe densities on either side of the boundary are d_1 and d_2 respectively then;

$$w = (1 - c)(d_1 + 1)(d_2 + 1)$$

6.3 A Graph To Represent Confidence Over The Fringe Field

Fig 8 shows a tiled section of the fringe field. Firstly each tile is considered to be a vertex in a weighted connected graph G (Fig 9). The graph is completed such that each tile vertex has edges, connected to the vertices which represent neighbouring tiles. That is, such that there is an edge (e) where two tiles have a common boundary. The weight (w) of this edge is calculated as described in section 3.2. Recap a low weighting indicates a good route and a high weighting a bad route.

The best path to take in assembling the tiles may then be found by constructing the minimum weight spanning tree of this connected undirected graph (Fig 10). This tree connects all of the tiles in the graph with optimum confidence. There are several known algorithms for the construction of this minimum spanning tree. Prim's algorithm has been utilised ref 11.

On each iteration of Prim's algorithm a new edge e is added to the growing spanning tree T. The edge e is the edge of least weight connecting a vertex in the remainder of the graph with those in T. The first vertex is selected such that it has the highest degree in the graph (maximum of 4) and the lowest sum of weights on its edges.

7 Results

7.1 Test Cases

The method has been applied to both ESPI (Electronic Speckle Pattern Interferometry) and Holographic fields. The results presented originate from three test cases, two ESPI and one Holographic. The form of the normalised input images, after preparation are shown in Figs 11, 15 and 19. These are respectively a board vibrating under the influence of an air jet (Holographic), a cylinder bore from a petrol engine (ESPI) and chamber (ESPI). The ESPI images were supplied by the Rover Group's Optics Laboratory at Gaydon (ref. 5).

The wrapped tan fringe fields for the above images are shown in Figs 12, 16 and 20. Fig 27 shows 3 highlighted areas of the chamber tan fringe field where difficulties have arisen. There are two tan fringes for each cosine fringe for the chamber. Images showing the low modulation noise are given in Figs 23, 24 and 25.

Normalised grey scale images representing the unwrapped phase map for the three cases are given in Figs 13, 17 and 21. It should be stressed that the output is unsmoothed. All filtering processes were completed at the generation of the wrapped phase map. The tile size in these images has been selected as 10 pixels by 10 pixels for the chamber, 20 by 20 pixels for the vibrating board and 40 by 40 pixels for the bore. Three dimensional mesh plots of the same data (unsmoothed) are given in figures 14, 18 and 22.

An example of how the tile size effects the solution is given in Fig 26. The grey scale height map for the chamber is shown with the tile size set at 30 by 30 pixels.

The last set of figures are tables showing the progression of each tree. Each box in the tables represents a tile, the number inside the box is an index to when the tile was added to the tree. The root of the tree is in the box labeled 1. Boxes which contain zero have failed to be processed. Fig 28 shows four tables, one for each quadrant of the vibrating board, the tile size here was 20. Fig 29 shows the much smaller tree for the bore, with a tile size of 40. Fig 30 shows the tree for the 30 pixel tile size solution for the chamber.

7.2 Discussion

The minimum spanning tree maximises confidence in the solution over the whole field, any approach that yields a tile solution may take advantage of such a tree.

The test fields considered are real and contain faults. The solutions show that they solve relatively well. There are however defects in the solutions. Two important points are that firstly these are traceable to deficiencies in the original input and secondly that the technique has minimised their affects.

The defects manifest themselves as steps at tile boundaries. They are caused when the fringe counts in following different branches of the tree to a tile do not match. Each field will be considered in turn.

The wrapped field for the chamber appears clear and unambiguous at first sight (Fig 20). However the solution shows three confused areas. The three regions are labeled A, B and C in Fig 27 with respect to the wrapped field. Region A shows an area where the bottom and top sections may be connected by crossing either two fringe boundaries (which may be thought of as contour lines) or none at all. There is no way to reconcile the two branches.

The algorithm follows the lower density route and the result is a sunken area in the solution. However, the spread of this area is contained. The branch which created it terminates as soon as there is an increase in fringe density. The solution tree then grows instead at some other point in the field where the fringe density is low. The same kind of argument may be applied in areas B and C.

In the centre of the chamber field there is a hole, caused by the spark plug. The tiles in this region of the field have failed because of a large number of low modulation points. The algorithm has processed around the hole.

The Holographic fringe field of the vibrating board is considered next (Figs 11 and 12). This field is interesting because the air jet has almost divided the field into two halves. The dividing line exhibits fine fringes and there is some ambiguity over this area. It may be seen that the fringes are least dense at the end of the air jet pipe, and so it is to be hoped that the connection of the halves will be made here.

The field has another problem in the lower left corner, near one of the nodes. It is caused by a fringe boundary fading out.

Both of these difficulties are seen in the tan fringe field (Fig 12). First of all consider the problem of connecting the two halves. The tree map illustrates the junction (Fig 28). The tree begins in box 1. Referring to the map for the right half of the field (top right and bottom right corner) it is seen that the transition has begun across the boundary of tiles 140 and 141. The connections continue in the right half 144-145-146-147-148. As was hoped these tiles are at the end of the air jet pipe. The fringe density then becomes too much for this branch of the tree, and the connections continue in the left side until a break point is reached with the connection of tile 172 to tile 147. The connections after this point are quickly traced up to tile 198 on the far right of the fringe field.

The affect of the faded fringe boundary, around tiles 40 to 60 is to create a small sunken region one fringe height below its surroundings. The size of this region is again limited by fringe density. As soon as the density of fringes increases the region is isolated. This can be seen just above the node, where the tile indices suddenly change from around the 50 mark to 100 or more.

The fringe field for the bore (Fig 16) shows a very marked trough just below and left of centre. Around this trough are fringes of gradually increasing height. However, close to the centre of the field the fringes become very badly disrupted. This is because the fringes were recorded from a conical mirror, at the centre of which was a hole, as a result the fringes around the centre are blurred. The outcome may be seen in the grey scale plot, near the centre, as a tile which is apparently unrelated to three of its neighbours.

7.3 The Affect Of Tile Size

In the solutions presented the tile sizes are 10, 20, 30 and 40 pixels. At a tile size of 10 pixels, with a 4 pixel overlap, there are 7506 tiles in a 512 by 512 pixel field. The calculation of fringe densities and edge confidences over this many tiles provides a vast amount of information upon which to base the tree construction. It is, perhaps, easier to understand how a good solution may be found with so much data than if the tile size is larger. The question arises, then, of whether the smallest tile size must always be used to give the best solution.

Let us inspect Fig 26 which is a solution for the chamber field where the tile size is 30 pixels by 30 pixels. As can be seen the field has been solved in a similar manner to the 10 by 10. It may be seen that in this solution the tiles have obvious striation marks. Large tiles are prone to these. They are caused when a fringe boundary is missed. The scan following the missed boundary falls out of sync until the edge of the tile. These marks are very short on a small tile and disrupt the field less. In the 10 by 10 solution it may be seen how well the tiles have fitted around the hole made by the spark plug, in the 30 by 30 solution the hole is square. It is clear that large tiles take with them more of the field if they fail than do small ones.

On the other hand the connection tree is constructed more rapidly with large tiles than with small. Perhaps more importantly there seems to be a correlation between the tile size which gives the best solution and the density of fringes, which might well be a reason for using larger tiles. A large tile has a greater overall view of the fringe field. Consider the chamber field again and the 30 by 30 solution tree (Fig 30). In the case of the 10 by 10 solution the area A has spread somewhat before the higher fringe densities of its surroundings isolate it (Fig 21). However, in the 30 by 30 solution (Fig 26), by virtue of the greater area each tile has in view a branch through the lower density spot is never taken, and the troublesome area is kept to a single tile (number 111). The solution in this area is therefore better with the larger tile. This subject needs more study.

8.0 Conclusions

A method for producing quasi-heterodyne holographic interferograms has been presented. In this case the two reference beam sources are close together in order to provide a spatially varying phase in the fringe image. A switching system was developed to allow one of two distinct reference beams to be selected for either of the pulses from a ruby laser. This system is electronically controlled and has a minimum pulse separation of 0.3 microseconds.

Hologram reconstruction was achieved using a single beam. A CCD camera and digital framestore provide the interface to a personal computer and subsequent image processing. Automation of the image capture system is possible using a computer controlled traverse.

The errors in the reconstruction of the holographic image have been estimated. If there is no correction of the phase map, the errors are nominally 1.75 μm over an object depth of 100 mm. However this can be reduced by subtraction of the imposed linear phase variation in the reconstructed image. This places the accuracy limitation on the pixel shift process, producing a phase error of 1 part in 19 for the particular holographic image shown.

The automation of a complete fringe analysis system becomes realistic with the advent of the automated phase unwrapping techniques presented. The minimum spanning tree provides a systematic method of producing a confident solution. The strength of the approach stems from the vast amount of information which tiling provides about the fringe field, such as local fringe density. The rejection of tiles containing bad data on the basis of both low modulation and lack of fringe consistency further improves confidence in the solution.

There is no execution time penalty in using the technique as the tree is formed in a single efficient pass for all but the smallest tile sizes. The result is a significant advance over more naive techniques which, because they do not segment the field into areas and sort on confidence, are unable to reject bad data and include its affects in the solutions they provide.

REFERENCES

1. R.Dandliker, R.Thalmann, "Heterodyne and Quasi-Heterodyne Holographic Interferometry", Optical Engineering, Vol.24, No.5, p.824-831 (1985)

2. P.Hariharan, "Quasi-Heterodyne Hologram Interferometry", Optical Engineering, Vol.24, No.4, p.632-638 (1985)

3. D.W.Robinson, "Automatic Fringe Analysis with a Computer Image Processing System", Applied Optics, Vol.22, No.14, p.2169-2176 (1983)

4. D.P.Towers, T.R.Judge, P.J.Bryanston-Cross, "Vibration Measurements Using Dual Reference Beam Holography", 2nd SIRA conference 'Stress and Vibration : Recent Developments', March 1989

5. J.Davies, C. Buckberry, "Applications of a Fibre Optic TV Holography System to the Study of Small Automotive Structures", SPIE proceedings 1163 (1989)

6. A.J.Decker, "Beam Modulation Methods in Quantitative and Flow Visualisation Holographic Interferometry", AGARD Conference Proceedings No. 399, pp.34-1 to 34-16 (1986)

7. G.Lai, T.Yatagai, "Dual-Reference Holographic Interferometry with a double Pulsed Laser", Applied Optics, Vol.27, No.18, p.3855-3858 (1988)

8. V.B.McKee, R.J.Parker, "A Numerical Simulation of the Quasi-Heterodyne Technique for Double Pulsed Holographic Vibration Measurements", 2nd SIRA conference 'Stress and Vibration : Recent Developments', March 1989

9. C.M.Vest, "Holographic Interferometry", p.68-72, Wiley Interscience series, New York (1979)

10. P.Hariharan, "Optical Holography", p.23-25, Cambridge University Press, (1984)

11. PRIM, R.C., "Shortest Connection Networks and Some Generalizations," Bell System Tech.J., Vol. 36,Nov. 1957,1389-1401.

12. DEO, N., "GRAPH THEORY with Applications to Engineering and Computer Science," Prentice-Hall, Inc. pp60 - pp64.

13. GIBBONS, A., "Algorithmic graph theory," Cambridge University Press. pp40 - pp41.

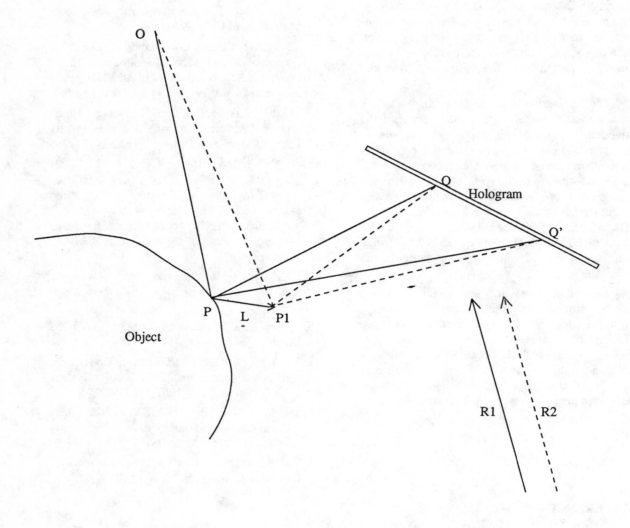

O - Illumination source
P - Initial position of a point on the object
P1 - Deformed position of point P
L - Vector translation of P
Q,Q' - Veiwing positions on the hologram
R1,R2 - Reference Beams

Fig.1 - Vector Diagram Representing Double
Exposure Holography

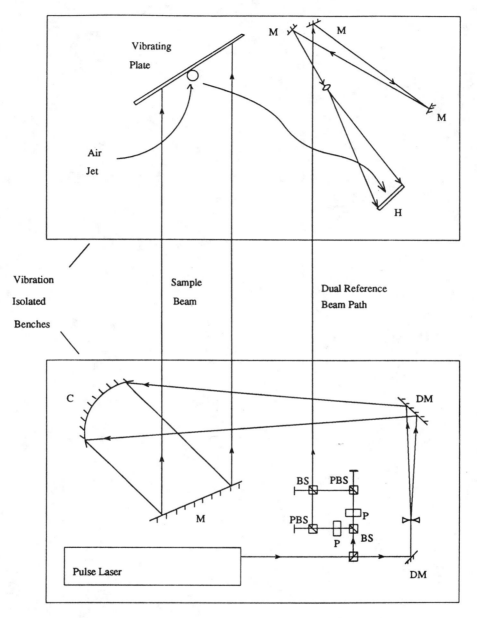

Vibrating
Plate

M   M

M

H

Air
Jet

Vibration   Sample   Dual Reference
Isolated   Beam   Beam Path
Benches

C   DM

BS   PBS

PBS
            P
                P
            BS
M   P

Pulse Laser   DM

<u>KEY</u>

PBS - Polarising cube beamsplitter   BS - 50/50 Beamsplitter

DM - Dielectrically coated mirror   P - Pockels cell

C - Collimator   M - Plane Mirror

H - Holographic Plate

Figure 2 - Holographic Arrangement

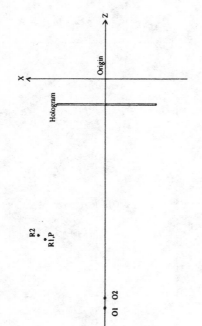

Hologram

X

Origin

Z

R2

R1,P

O1 O2

R1,R2 - Two Reference Beam Sources

P - Reconstruction Beam Source

O1,O2 - Sample Object Posistions For Pulses One And Two

Figure 4 - Definitions For Error Analysis

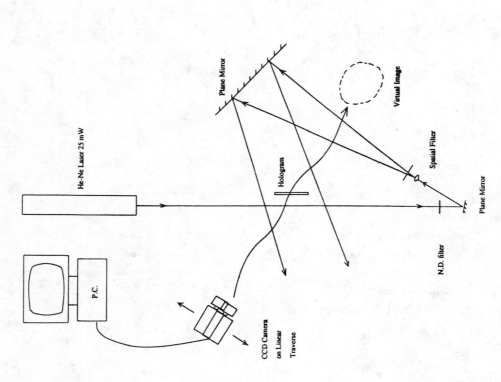

He-Ne Laser 25 mW

Plane Mirror

Virtual Image

Hologram

Spatial Filter

P.C.

CCD Camera
on Linear
Traverse

N.D. filter

Plane Mirror

Figure 3 - Reconstruction Geometry

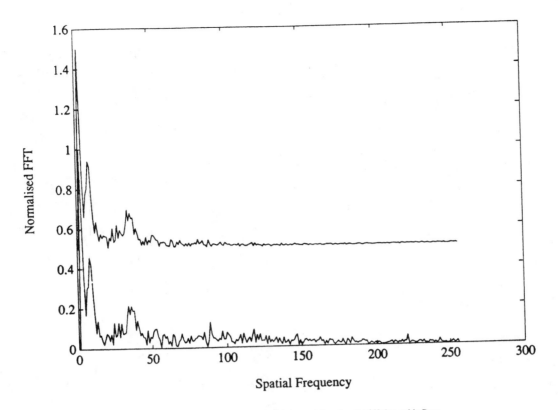

Figure 5a - Raw and Averaged Spectrums of Holographic Data

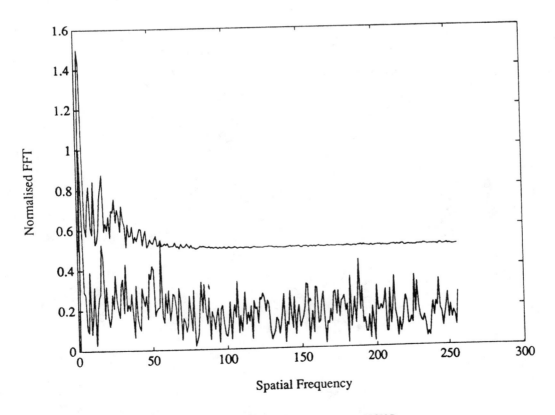

Figure 5b - Raw and Averaged Spectrums of ESPI Data

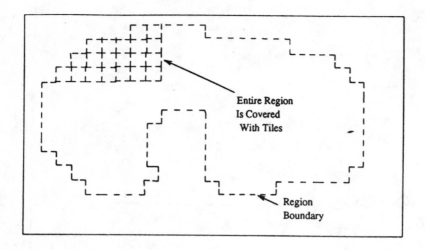

Fig 6 Boundary definition and tile coverage.

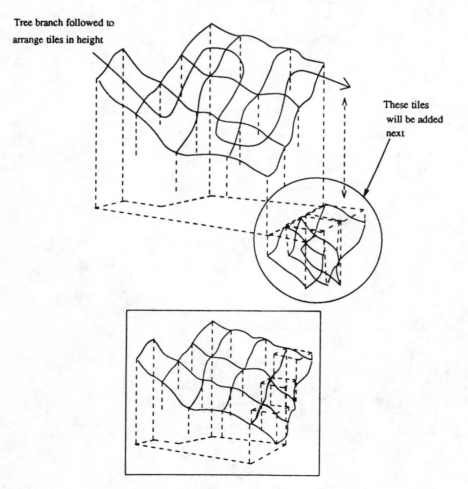

Fig 7 Tiles being arranged at their correct height offsets.

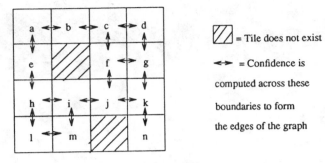

= Tile does not exist

⟷ = Confidence is computed across these boundaries to form the edges of the graph

Fig 8 Tiled section of the fringe field.

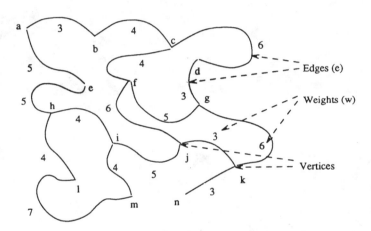

Edges (e)

Weights (w)

Vertices

Fig 9 Graph G constructed from the tiled section showing confidence levels

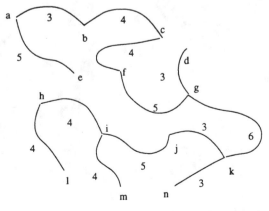

Fig 10 The minimum weight spanning tree T giving optimum confidence

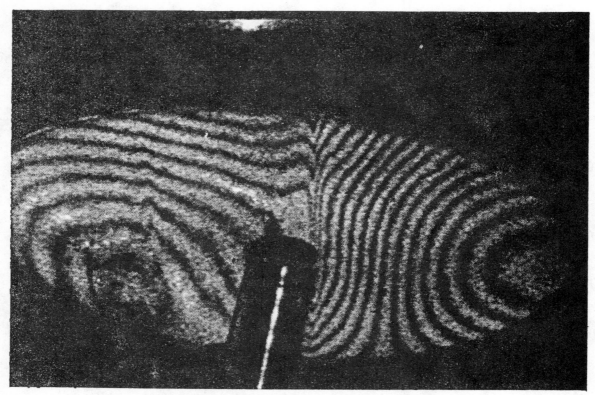

Fig 11 Holographic Cosine Fringe Field For Vibrating Board

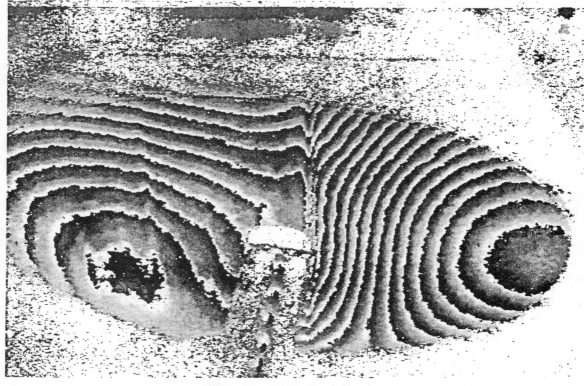

Fig 12 Tan Fringe Field For Vibrating Board

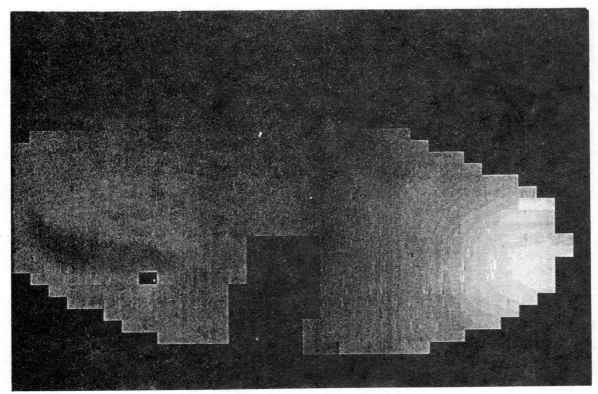

Fig 13 Grey Scale Height Map For Vibrating Board (20 by 20)

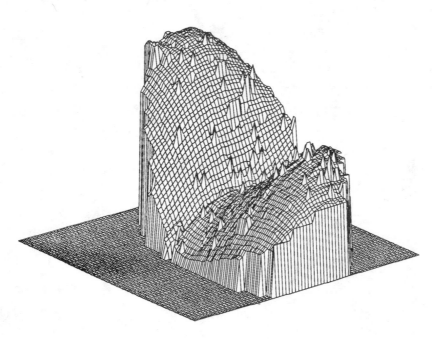

Fig 14 Mesh Plot Of Height Map For Vibrating Board

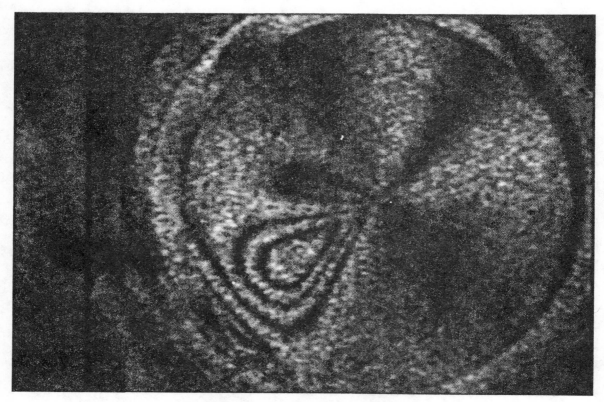

Fig 15 ESPI Cosine Fringe Field For Bore (After Preparation)

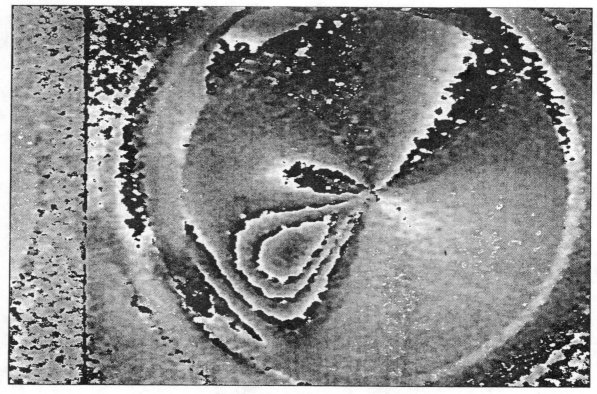

Fig 16 Tan Fringe Field For Bore

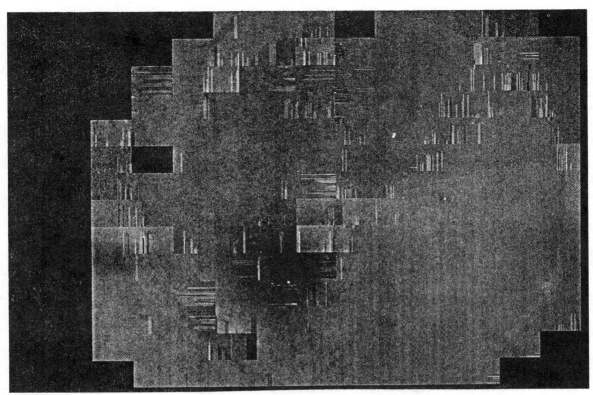

Fig 17 Grey Scale Height Map For Bore (40 by 40)

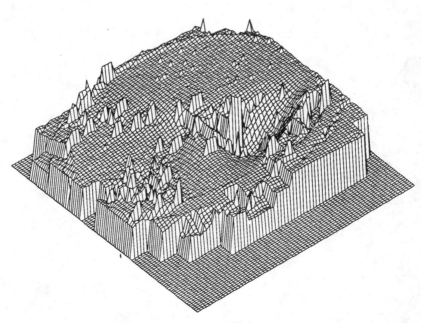

Fig 18 Mesh Plot Of Height Map For Bore

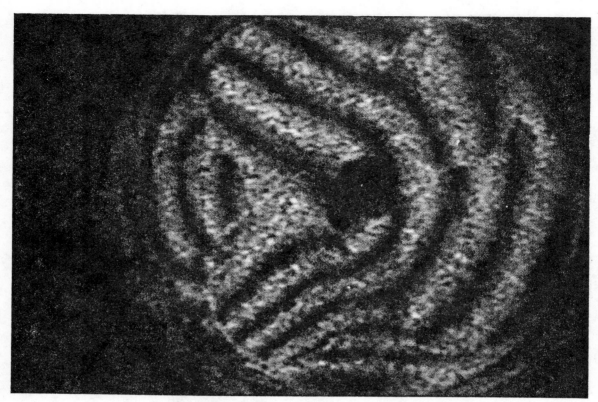

Fig 19 ESPI Cosine Fringe Field For Chamber (After Preparation)

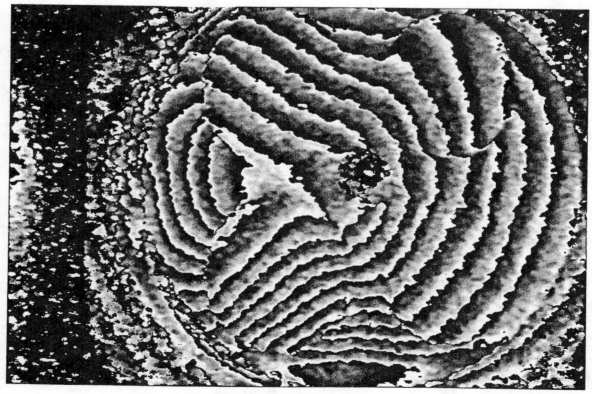

Fig 20 Tan Fringe Field For Chamber

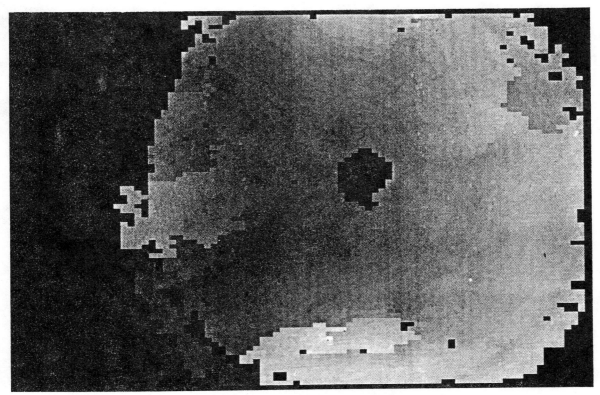

Fig 21 Grey Scale Height Map For Chamber (10 by 10)

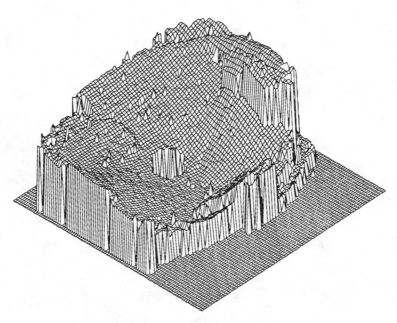

Fig 22 Mesh Plot Of Height Map For Chamber

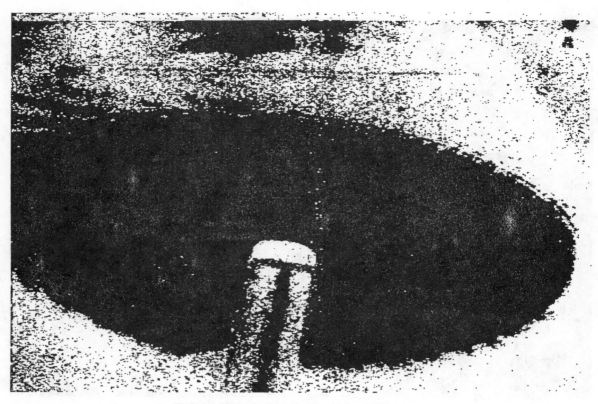

Fig 23 Low Modulation Points For Vibrating Board (White Points)

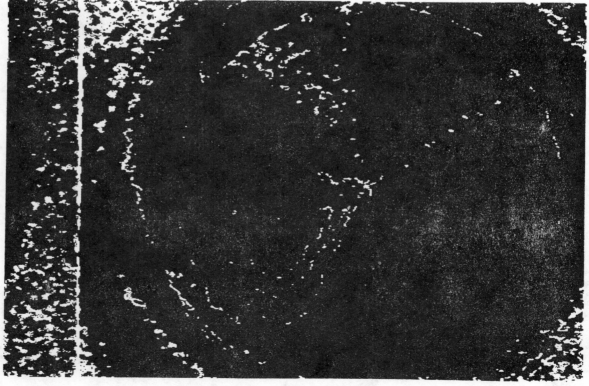

Fig 24 Low Modulation Points For Bore (White Points)

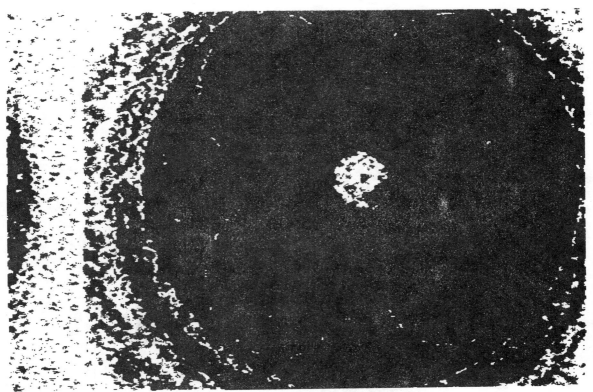

Fig 25 Low Modulation Points For Chamber (White Points)

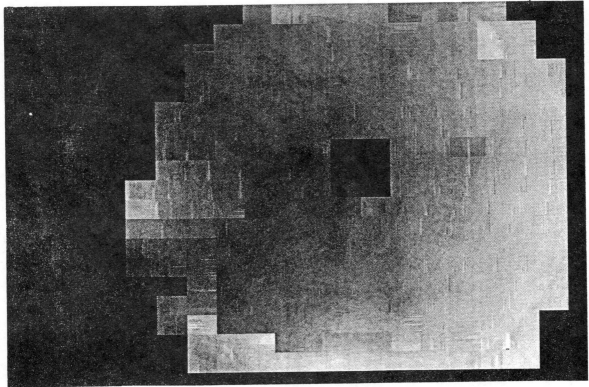

Fig 26 Grey Scale Height Map For Chamber (30 by 30)

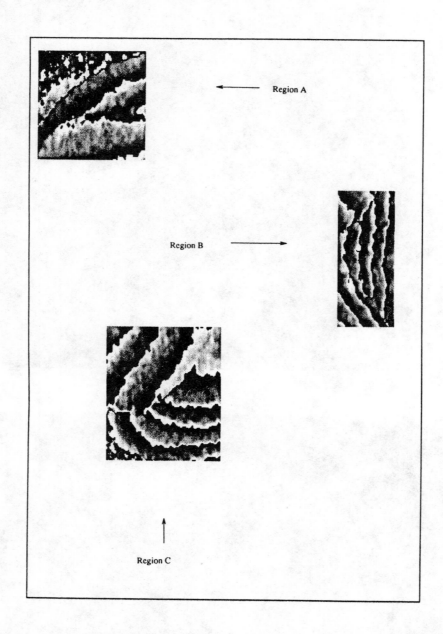

Fig 27 Troublesome Areas In Tan Fringe Of Chamber

0	0	0	0	0	0	0	0	0	0	0	0	0	0	0
0	0	0	0	0	0	0	0	0	0	0	0	0	0	0
0	0	0	0	0	0	0	0	0	0	0	0	0	0	0
0	0	0	0	0	0	0	0	0	0	0	0	0	0	0
0	0	0	0	0	0	0	0	0	0	0	0	0	0	0
0	0	0	0	0	0	0	0	0	0	0	0	0	0	0
0	0	0	0	0	0	0	0	0	0	0	0	0	0	0
0	0	0	0	0	0	0	0	0	0	0	0	0	0	0
0	0	0	0	387	385	386	414	415	418	421	358	357	422	360
0	379	381	382	383	384	388	419	416	417	449	354	351	352	353
380	378	377	376	375	374	367	347	346	345	348	349	350	394	355
338	339	109	389	390	391	392	393	373	204	205	206	207	170	306
102	103	104	105	110	111	112	290	209	203	208	155	154	151	152
101	115	119	106	107	108	117	364	365	133	136	137	280	150	167

100	97	98	99	129	128	125	126	127	130	131	132	138	139	140
95	96	93	74	75	81	122	86	135	134	166	165	159	142	143
53	54	69	70	71	76	83	84	85	120	121	163	160	161	158
51	52	73	72	94	90	91	92	334	335	169	168	162	164	149
49	50	55	56	38	37	34	33	35	48	57	58	77	0	0
44	43	42	41	39	40	36	32	31	30	28	29	118	0	0
45	46	60	47	78	79	116	80	82	59	26	27	87	0	0
0	62	61	68	318	317	319	0	99	23	24	25	88	0	0
0	0	63	64	65	321	320	322	19	18	114	113	0	0	0
0	0	0	66	67	20	336	307	123	15	21	22	0	0	0
0	0	0	0	0	12	11	13	9	14	16	124	0	0	0
0	0	0	0	0	0	0	4	3	1	2	10	17	0	0
0	0	0	0	0	0	0	0	5	6	7	8	0	0	0
0	0	0	0	0	0	0	0	0	0	0	0	0	0	0
0	0	0	0	0	0	0	0	0	0	0	0	0	0	0

0	0	0	0	0	0	0	0	0	0	0	0	0	0	0	0
0	0	0	0	0	0	0	0	0	0	0	0	0	0	0	0
0	0	0	0	0	0	0	0	0	0	0	0	0	0	0	0
0	0	0	0	0	0	0	0	0	0	0	0	0	0	0	0
0	0	0	0	0	0	0	0	0	0	0	0	0	0	0	0
0	0	0	0	0	0	0	0	0	0	0	0	0	0	0	0
0	0	0	0	0	0	0	0	0	0	0	0	0	0	0	0
451	0	456	0	0	0	0	0	0	0	0	0	0	0	0	0
359	363	398	441	440	431	439	0	0	0	0	0	0	0	0	0
356	361	362	435	429	430	434	433	0	0	0	0	0	0	0	0
202	399	400	436	426	427	428	432	437	438	0	0	0	0	0	0
153	157	458	448	423	443	455	454	457	450	0	0	0	0	0	0
156	424	425	412	402	401	442	403	326	372	408	409	410	0	0	0

141	171	453	289	329	330	331	327	325	444	445	316	396	397	0	0	
144	371	173	199	200	405	293	323	324	368	314	313	315	342	341	0	
145	147	172	174	175	183	182	340	201	192	191	193	210	332	333	0	
146	148	292	181	176	177	178	184	185	186	190	194	195	196	198	0	
0	0	291	211	288	180	179	395	406	187	188	189	197	268	266	267	
0	0	452	212	277	224	328	411	279	261	260	274	269	250	251	252	
0	0	370	213	214	215	216	276	287	255	253	249	248	247	246	0	
0	0	420	366	446	217	218	219	257	256	254	262	244	243	245	0	
0	0	447	278	221	220	222	223	263	259	258	240	239	241	242	0	
0	0	407	227	225	226	264	233	234	235	236	237	238	265	0	0	
0	0	404	229	228	230	231	232	273	272	271	270	312	0	0	0	
0	283	282	281	275	294	303	302	308	309	310	311	0	0	0	0	
0	284	337	369	304	295	298	301	343	344	0	0	0	0	0	0	
0	285	286	305	297	296	299	300	0	0	0	0	0	0	0	0	
0	0	0	0	0	0	0	0	0	0	0	0	0	0	0	0	
0	0	0	0	0	0	0	0	0	0	0	0	0	0	0	0	

Fig 28 Minimum Weight Spanning Tree For Vibrating Board

Whole Field For Bore With Tile Size Of 40 By 40 Pixels

0	0	0	0	0	133	134	136	0	86	77	145	146	0
0	0	0	0	115	132	135	149	140	80	76	79	148	88
0	0	0	119	113	114	108	144	129	78	75	81	84	85
0	0	0	118	106	105	107	143	121	82	74	73	72	137
0	0	131	109	101	102	103	110	111	83	87	70	61	62
0	0	154	0	100	99	104	112	120	122	90	58	59	71
0	0	147	95	94	96	98	126	123	65	64	48	33	34
0	0	117	116	93	97	130	151	66	32	29	28	27	26
0	0	56	57	92	127	128	139	36	21	22	23	24	25
0	0	54	55	67	89	141	142	60	20	19	18	17	30
0	0	52	53	125	153	152	124	11	10	12	13	14	15
0	0	51	50	138	150	69	68	37	9	8	7	16	63
0	0	44	43	45	91	47	46	2	1	5	6	38	0
0	0	0	42	41	40	39	31	3	4	35	49	0	0

Fig 29 Minimum Weight Spanning Tree For Bore

Whole Field For Chamber With Tile Size Of 30 By 30 Pixels

0	0	0	0	0	0	0	0	90	112	78	68	67	66	176	60	61	0	0
0	0	0	0	0	0	0	76	89	91	192	73	77	65	54	55	88	87	0
0	0	0	0	0	0	199	75	74	72	71	70	69	64	52	53	111	86	0
0	0	0	0	0	0	80	79	84	85	63	58	57	56	47	46	44	45	217
0	0	0	0	0	0	95	94	93	92	62	51	50	49	48	59	43	42	171
0	0	0	0	0	100	98	96	97	107	106	105	104	103	101	41	38	39	40
0	0	0	0	0	108	166	99	21	20	22	23	31	32	33	34	35	36	37
0	0	0	0	0	109	110	183	10	7	8	0	0	128	131	163	81	82	83
0	0	0	0	0	30	29	3	1	2	9	0	0	123	115	114	113	102	162
0	0	0	0	175	25	19	18	4	5	6	0	0	124	121	116	117	119	120
0	0	0	0	173	172	174	244	12	11	13	186	126	125	127	129	118	122	130
0	0	0	0	182	181	180	16	15	14	184	185	193	177	134	133	132	135	136
0	0	0	0	231	28	27	26	17	24	216	195	179	178	187	164	149	142	137
0	0	0	0	197	196	198	188	189	194	229	169	168	167	161	157	146	145	138
0	0	0	0	0	0	222	190	191	219	226	225	224	160	159	158	150	140	139
0	0	0	0	0	234	223	221	232	233	242	227	155	154	153	152	147	141	148
0	0	0	0	0	235	246	240	241	236	237	243	230	228	156	151	144	143	0
0	0	0	0	0	0	215	245	211	239	238	220	212	247	170	205	218	165	0
0	0	0	0	0	0	214	213	210	209	208	207	206	203	202	204	201	200	0

Fig 30 Minimum Weight Spanning Tree For Chamber (30 by 30)

CCD Based Moire Interferometric Strain Sensor (MISS)

A. Asundi and Kan Man Fung

Department of Mechanical Engineering
University of Hong Kong
Hong Kong

ABSTRACT

The Moire Interferometric Strain Sensor (MISS) is based on the principle that gratings diffract light in preferred directions determined by their frequency. Hence by tracking the change in diffraction angle the frequency of the grating can be determined. In engineering applications, a grating attached to the specimen would change frequency as a consequence of the strain at that point. Thus the change in diffraction angle can directly be related to strain. Since for most elastic strain measurements the change in diffraction angle is very small, this paper describes the use of a CCD array to improve the sensitivity.

2. INTRODUCTION

Gratings diffract light in predetermined directions depending on their frequency. If the frequency of the grating changes, then the diffraction angle would change accordingly. Moire interferometry[1] uses diffraction gratings to monitor the deformation of the object. In this method, a high frequency grating is generated on the specimen surface and illuminated by two symmetric beams of coherent light. Each beam is diffracted by the specimen and a pair of diffraction orders emerge normal to the grating plane. Deformation of the specimen results in change of specimen grating frequency; which in turn deviate the diffracted beams. A two beam interference pattern is generated which contours the in-plane displacement component.

In this paper, the change in frequency, which is a direct measure of the strain, is determined by measuring the deviation in the diffraction angle. For small changes in frequency, this deviation is small and thus a CCD array is used for increased sensitivity and accuracy.

3. THEORY

Consider the experimental schematic, shown in Fig. 1. A narrow beam of laser light illuminates a point on a high frequency grating at an angle 'α'. The grating diffracts the incident beam into preferred directions according to the diffraction equation

$$\sin \beta = \sin\alpha + m\lambda f \qquad (1)$$

where β is the diffraction angle, m is the order of diffracted beam, λ is the wavelength of the laser and f is the frequency of the grating.

Initially 'α' is chosen such that the (-1) diffraction order emerges normal to the grating plane i.e. $\beta = 0°$. When the specimen deforms, the grating frequency changes and correspondingly the diffracted beam no longer emerges normal to the grating plane. Differentiating eqn. (1) we get

$$\cos\beta \; d\beta = -\lambda df$$

$$\text{i.e.} \; d\beta = -\lambda df \tag{2}$$

$$\text{since} \; \beta = 0°$$

Thus the deviation of the diffracted beam is directly proportional to the change in frequency. Now change in frequency is equal to the initial frequency times normal strain assuming that deviation of the diffracted beam is along the principal direction of the grating. Thus

$$d\beta = \lambda f \varepsilon_x \tag{3}$$

If the shift of the diffraction dot (δ) is observed on a screen placed at a distance 'L' from the grating then

$$d\beta = \frac{\delta}{L} = \lambda f \varepsilon_x \tag{4}$$

$$\text{i.e.} \; \varepsilon_x = \frac{\delta}{L} \cdot \frac{1}{\lambda f}$$

In general the deviation of the diffracted beam is in two dimensions. Thus the diffracted beam would shift in both the x and y directions (Fig. 1). The shift in the x-direction would give the normal strain according to eqn. (4), while the shift in the y-direction is representative of the gradient $\partial u/\partial y$.

4. EXPERIMENT

As an initial demonstration of the MISS, an experiment involving rigid body rotation was performed. A 600 lines/mm grating was mounted on fixture capable of in-plane rotation of the grating. A dial gauge was used to measure the deflection and thus the rotation. The MISS system was similar to Fig. 1, except that the shift was measured on a screen. The distance of the screen from the grating was 0.533 m, Fig. 2 shows the shift of the diffraction dot corresponding to a rotation of 0.031 rad. The shift is predominantly in the vertical direction as expected and a comparison of the rotation as measured using MISS and using the dial gauge provides confirmation of eqn. (4). However, there is also a small shift in the horizontal direction. This is due to out-of-plane tilt of the specimen.

This can be easily avoided by using two beams as in moire interferometry. If the two beams are incident along '±α' directions, then only a change in frequency would cause relative movement between the diffraction dots. Another way of doing this would be to observe the relative movement of diffraction dots from two neighbouring points on the specimen. In this case only strain induced change in pitch would result in relative movement of dots, while out-of-plane tilt and rotation would maintain the same relative spacing before and after. An example of this is shown in Fig. 3 which depicts the movements of diffraction dots on a specimen subject to rigid body rotation as in the earlier example. The lower pair of dots are from two points 30 mm apart on the specimen before rotation, while the upper pair are recorded after a rotation of 0.04 rad. The screen was at a distance of 1.3 m from the grating resulting in the much larger shift as compared to Fig. 2. However, as expected there is no relative shift between the diffraction dots after the rotation, indicating that the frequency of grating before and after rotation is the same at the two points. In other words, there is no relative strain between the two points. Fig. 4 shows the effect of relative strain on the relative shift of the diffraction dots. The specimen in this case was a quasi-isotropic graphite epoxy composite tensile coupon with a central hole. The upper pair of dots are the initial diffraction at a distance of 1 m, for two points on the specimen - one point (the diffraction dot to the right) was adjacent to the hole and the other 11 mm further away. After the specimen was loaded to a far-field strain of 0.0006, the diffraction dots were spaced as in the lower pair. The relative strain between the two points could be determined by measuring the change in spacing of the diffraction dots. This was, in this instance, 0.0017. The individual strains could also be obtained at these points by measuring the shift of the individual dots before and after loading.

In the above examples, the shift was measured mechanically and thus the screen had to be placed at a large distance from the grating and the rotation and strains had to be fairly large. For better measurement accuracy, the screen was replaced by a CCD array made up of 450,000 pixels. The signal from the CCD was fed into a frame grabber board in a IBM-AT compatible - personal computer. Relative shifts could then be tracked on the display monitor and addition of pseudo-colour to the display enhanced the measurement accuracy. Fig. 5 shows the black and white version of the diffraction dots recorded from the colour monitor. The specimen was the same composite specimen as in Fig. 4 except that the point under study was 19 mm away from the hole. The corresponding far-field strains from top to bottom were 100, 200 and 360 $\mu\epsilon$ respectively. The CCD distance from the specimen was large 1670 mm. However much smaller strains resulted in larger shifts, thus increasing the accuracy. The two beam arrangement could also be extended for this case. However since the CCD array has a limited size, it is desirable that initially the two beams from two points on the specimen overlap. Increasing strain would cause the two beams to separate while out-of-plane tilt or rigid body rotation would not. An example for the composite specimen is shown in Fig. 6. As before these are black and white reproductions of the colour photographs recorded off the monitor. The top photograph is the initial pattern, while the next two correspond to far-field strains of 220 and 360 $\mu\epsilon$ respectively. The sensitivity for measurement in this case is 15 $\mu\epsilon$/column of shift on the monitor. This is already much improved as compared to pure mechanical measurements and the recording distance can be significantly reduced while maintaining fairly good sensitivity.

5. CONCLUSION

A moire interferometric strain sensor has been developed. The sensor measures changes in frequency of the specimen grating to determine the strain to improve the sensitivity of the sensor a CCD array is used to monitor the shift of the diffraction dot. The sensitivity can be adjustable by varying the distance between the array sensor and the specimen grating. Furthermore by simultaneous interrogation of two neighbouring points, shifts due to rigid body rotation or out-of-plane tilt can be avoided.

6. REFERENCES

1. D. Post, "Moire Interferometry" in Handbook on Experimental Mechanics ed. A.S. Kobayashi, pp. Pergamon Press 1987.

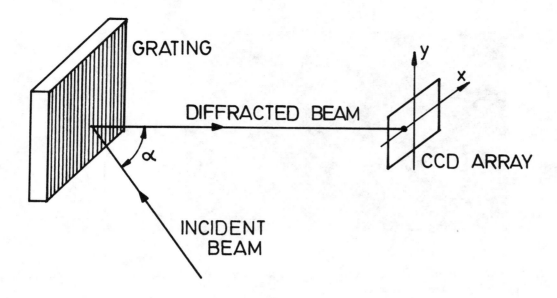

Fig. 1   Experimental schematic of MISS system.

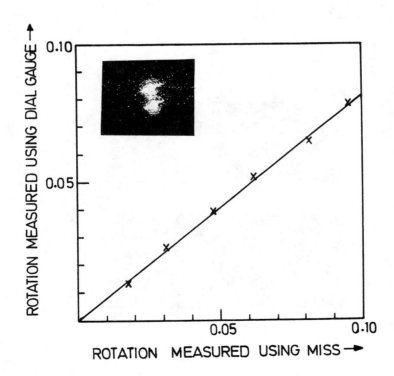

Fig. 2   Experimental corroboration of MISS.   Inset shows a typical shift of
diffraction dot due to rigid body rotation.

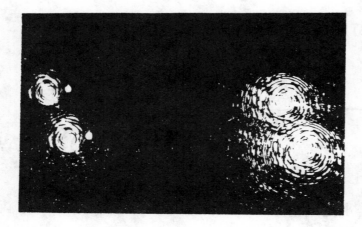

Fig. 3   Shift of diffraction dot from two neighbouring points on object grating
due to rigid body rotation.

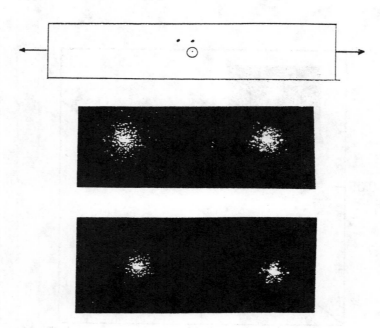

Fig. 4   Shift of diffraction dot from two neighbouring points on the object due
to strain.

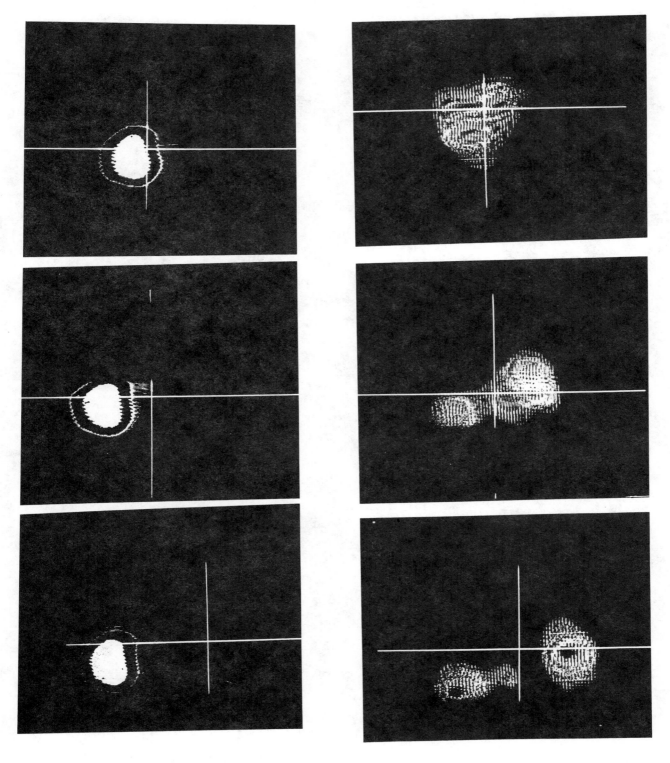

Fig. 5   Successive frames showing shift
         of pseudo-colour enhanced
         diffraction dot recorded off a
         colour monitor.

Fig. 6   Two beam arrangement showing
         shift due to relative strain.

SESSION 3

Applications and Studies of Phase Measurement

Chair
Suezou Nakadate
Tokyo Institute of Polytechnics (Japan)

Profile measurement by projecting phase-shifted interference fringes

S.Kakunai, K.Iwata[*], M.Hasegawa and T.Sakamoto

Himeji Institute of Technology, Department of Mechanical Engineering,
Shosha 2167, Himeji, Hyogo, 671-22, Japan
[*]University of Osaka Prefecture, Department of Mechanical Engineering,
Mozu Ume, Sakai, Osaka, 591, Japan

Abstract

Profile measurement using phase-shifted interference fringe projection technique is described and discussed. This method realizes the measurement by projecting interference fringe patterns on the object surface and by observing the deformed fringe pattern at the direction different from the projection. To obtain the information finer than the fringe spacing, phase-shifting technique is adopted. This technique needs no fine fringe pattern to improve lateral resolving power as it gives height distribution at all the picture elements over the object image. Discussions are made on errors inherent to the method, including the error depending on intensity quantization and number of phase-shifting steps.

1. Introduction

Recently, a variety of fields requires non-contact, accurate and speedy profile measurement of object surface. Various methods have been proposed[1~4], which can measure the profile over the whole surface at an instant. Grating projection method consists of a relatively simple optical system. The method realizes the measurement by projecting a grating pattern on the object surface and by observing the deformed grating pattern at the direction different from the projection. In the conventional fringe pattern analysis, the data for the profile measurement is obtained only at the peak position of the bright fringe. However, detecting the peak position is a difficult problem for computer processing.

Several techniques to solve these problems have been proposed. Among them, Fourier-tansform technique[1] based on the Fourier analysis of spatial intensity variation of the projected grating. It realized automatic measurement with the lateral resolving power of less than a rating pitch. But, when there is spatial rapid variation of reflectivety on the object surface, the variation cannot be distinguished from the intensity variation of the grating image. Thus, this method cannot be applied to objects with fine letters on them. To separate the unwanted disturbance, we have to use finer gratings.

Phase-shifting technique[5,6] is based on the Fourier analysis of sequential intensity variation of the projected grating or interference fringes. It requires more than three different images and takes more time compared with the above Fourier-tranform technique. This technique, however, needs no fine fringe pattern to improve lateral resolving power as it gives height distribution at all the picture elements over the object image. Furthermore, it may be basically possible to measure the object with spatially varying reflectance such as letter or figure on the object surface.

This paper takes up the phase-shifted interference fringe projection technique[7,8] discusses the errors inherent to the method, including the error depending on the intensity quantization and number of phase-shifting steps.

2. Measuring principle

The optical system geometry for explaining the principle of the present method is shown in Fig.1. In the figure, point A is the projection center of interference fringe pattern and point O is the center of the imaging lens of the object. The z axis is taken in the direction of optical axis of imaging system. Equi-intensity positions of the projected fringes are assumed to form planes like m in Fig.1 and all the planes are perpendicular to the x-z plane. P is a point on the object surface and P' is the foot of the perpendicular from P to the x-z plane. If the angle α, ξ for the point assumed P' can be measured, the distance Z_0 from the x-y plane for the point P can be calculated from the equation:

$$Z_0 = a/\{ \tan \xi + \tan \alpha \} . \tag{1}$$

The angle ξ can be obtained by the coordinates of the image point P in the image sensor. For example, in the optical system with no lens distortion, the angle ξ is given by

$$X = Z \tan \xi . \tag{2}$$

To obtain the angle α, a known interference fringe pattern expanding from point A is projected on the object surface. Now suppose that the bright plane of the projected fringe is the plane m in Fig.1. The angle α can be obtained from bright line on the object image by knowing in advance the value of α for bright line.

In the conventional grating projection method, the angle α is caluculated at the peak position of bright fringe. Thus, the profile of the narrow area which is smaller than fringe pitch can not be measured.

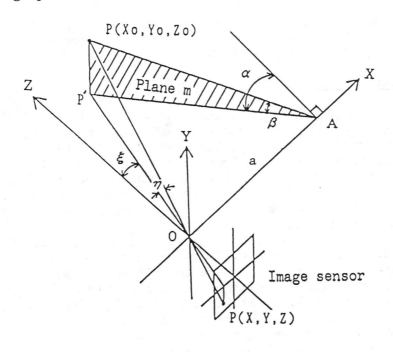

Fig.1 Optical system geometry for
the present method

In the present method, an interference fringe pattern as a grating pattern is projected on the object surface. The intensity I_0 of interference fringe is given by

$$I_0(\alpha,\beta,r) = I'(\alpha,\beta,r)\{ 1+ \gamma(\alpha,\beta,r) \cos(\phi(\alpha)+\phi_0)\} , \qquad (3)$$

where ϕ_0 is the phase shift given to obtain a more accurate value of the phase $\phi(\alpha)$, r is the distance from the projecting point A, $I'(\alpha,\beta,r)$ is the mean intensity and $\gamma(\alpha,\beta,r)$ is the visibility in the interference fringe pattern. In the equation, it is assumed that the fringe intensity depends only on the angle α.

The projected interference fringe is scattered by the object surface and imaged on an image sensor. The intensity of the object image is changed by spatial variations such as the reflectance, the gradient of the object surface and the imaging optical system. If these effect is denoted by $R(x,y)$, we can write the intensity I on the image sensor

$$I(x,y) = R(x,y)I_0(\alpha,\beta,\gamma) . \qquad (4)$$

In the phase-shifting technique, we make three or more measurements of the intensity distribution for different phase ϕ_0 and calculate the phase $\phi(\alpha)$ from the obtained intensity I_i. If the intensity is measured for N steps of phase ϕ_0, the phase $\phi(\alpha)$ can be calculated from the equation

$$\phi(\alpha) = \tan^{-1} \{ \Sigma\, I_i \sin(2\pi i/N)/ \Sigma I_i \cos(2\pi i/N)\}, \quad (i=1,2, \quad N). \quad (5)$$

In eq.(5), the phase $\phi(\alpha)$ gives principal value of the arctangent ranging from $-\pi/2$ to $\pi/2$. The discontinuity of phase profile occurs when $\phi(\alpha)$ is equal to the integral multiple of $\pm\pi$. We have to correct the discontinuity by assuming that the profile is smooth and continuous.

We have to note the $I'(\alpha,\beta,r)$, $\gamma(\alpha,\beta,r)$ and $R(x,y)$ does not affect the resultant phase $\phi(\alpha)$. Furthermore, the phase $\phi(\alpha)$ can be obtained at the every picture element irrespective of the projected fringe spacing.

3. Experimental system and verification experiment

The experimental optical system is shown schematically in Fig.2. The fringe pattern projected on the object surface is made of a compact fringe forming interferometer. It consists of a beam splitter and two mirrors, one which is mounted on a piezoelectric translater. The beam from an argon ion laser is split in two beams by a beam splitter. The interference fringe is expanded by the objective lens and projected on the object surface. The fringe spacing can be changed easily by tilting of the mirror. The deformed fringe pattern on the object surface is observed with an image sensor at the direction different from the projection.

The fringe pattern is digitaized at 512x512 picture elements with an intensity depth of 256 gray levels by CCD camera. The digitaized image is stored in digital frame memory in 1/30 sec.. The fringe pattern is shifted by applying a voltage to the P.Z.T device attached to a mirror. A microcomputer provides the required control signals to the camera, the P.Z.T. and the frame memory. The stored data can be transferred into the microcomputer's own memory for processing.

The errors such as each geometric parameter of the optical system and aberrations of the lenses of the CCD camera are compensated as a system error in advance by measurement of a standard plane surface.

To verify the method, the measurement of cylindrical surface of radius 41 mm was performed. In this experimental optical arrangement, object surface is placed at the distance 672 mm from the center of the imaging lens of the object on the Z axis as shown in Fig.2. The distance a between O and A is also 672 mm. The phase is calculated from eq.(5) using the four-step method of $\pi/2$, π, $3\pi/2$ and 2π. As shown in Fig.3(a), experimental value agrees well with the ideal cylindrical shape. Fig.3(b) shows the measurement error distribution of the cylindrical shape for the ideal one. The error is less than ±0.3 mm.

A 3-D diffuse object shape, a tooth model which is four times as large as a tooth was also measured. Fig.4(a) shows the deformed fringe pattern produced by this object. Fig.4(b) is a graphical representation of the result. In this figure, 14 profile lines with about 55 points on each line are shown.

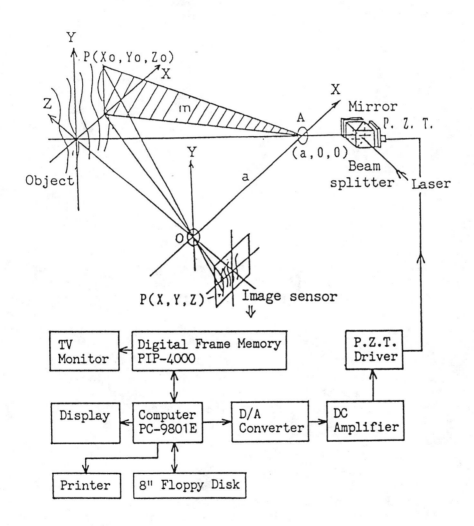

Fig.2 Schematic system of the experimental setup

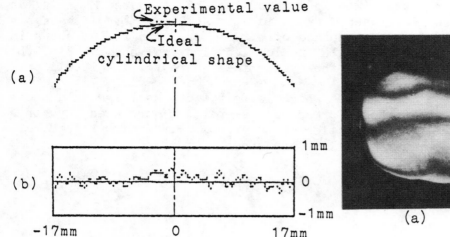

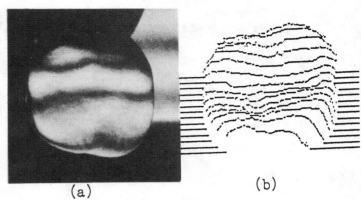

Fig.3 Experimental results of a
cylindrical shape
(Projection pitch = 11.3 mm)
(a) Comparison of the experimental
results with an ideal
cylindrical shape
(b) Measurement error distribution
of a cylindrical shape

Fig.4 Profile measurement of a tooth model
(Projection pitch = 17.3 mm)
(a) Deformed fringe pattern
(b) Sectional shapes calculated by
the phase-shifting technique

4. Discussion on error and range

4.1 Phase error

Accuracy of angle α is determined by the accuracy of phase determined by eq.(5).
The latter is influenced by a variety of properties of the actual system. They
include phase-shifting error, nonlinearity of the detector, quantization of the
detected intensity and number of phase-shifting steps. Sume of these error sources
has been investigated for phase shifting interferometry[6]. Here similar simulations
are performed considering that phase shifting tecnique is used for profilometry using
interference fringe projection.

Phase-shifting error is the error caused by the inaccurate calibration of the
phase shifting device such as the piezoelectric transducer in the above experiment.
The proper phase shift for the N phase shifting system is $2\pi/N$. In the simulation,
actual phase shift ϕ_0 is assumed to be expressed as

$$\phi_0 = 2\pi/N \, (1 + \epsilon) \, . \qquad \qquad (6)$$

Quantization level is determined by the A/D converter and the contrast of the
detected fringes. In the usual image processing system, the full range of the A/D
converter is digitized into 8 bit. But we cannot use the full range because the
fringe contrast is not unity all over the object surface, Moreover the intensity of
the scattered light is not uniform over the surface. The non-uniformity of intensity
is larger for the fringe projection profilometry than for the phase shifting
interferometry because the object to be measured is a rough one, whose surface may be
inclined to the viewing direction and may have varying reflectance.

The number of phase shifting steps is N in eq.(5). In the above experiment, four-step system is adopted. If we consider the speed of image processing, fewer step system is preferred. Present simulation investigates improvement of phase error for the larger step systems.

Nonlinearity of the detector may influence the phase error, but in this simulation it is not taken into consideration. As the detected intensity varies over the surface in the fringe projection profilometry, fringe contrast may be low and so the nonlinear effect may be small.

In the simulation, we first assume a known phase, calculate the intensity for N-step phase shift with phase shift error, quantize the intensity and calculate the phase using eq.(5). The difference between the calculated phase and the assumed phase is considered as error. The error is calculated for the 100 known phases ranging from $-\pi/2$ to $\pi/2$. Phase errors like Fig.5 is obtained which change with the assumed phase. As a measure $\Delta\phi$ of the over all phase error we adopt the root mean square error for the 100 errors. The results of the simulation is shown in Fig.6.

The result shows that error $\Delta\phi$ decreases as the number of quantization level increases when no phase shift error is present. But the error does not become smaller than a certain value when phase shift error is present. If the phase shift error ε is 1 percent, quantization level finer than 5bit does not seem effective as far as the phase error is concerned. Increase in number of phase shift steps generally decreases the error but the effect is not so considerable. We should note that even if the quantization level is binary, the phase error can be smaller than 5 deg for large N.

4.2 Height error

The phase obtained by the phase shift method is converted into the projection angle α. If the spacing of the projected fringes is smaller, error in α becomes proportionally smaller for a same phase error. Thus if we want accurate height, we have to use fine fringes.

The height is calculated from ξ and α using eq.(1). ξ in turn is obtained from the position of the picture element. The correspondence between the picture element and the angle is determined by simple geometry. It can be calibrated in advance using a standard test chart. we consider here that the calibration can be made perfect.

In this conditon, error of α, $\Delta\alpha$ is converted into height error ΔZo with eq.(1).

$$\Delta Zo = a \ \Delta\alpha \ / \ \{ \ (\ \tan \xi + \tan \alpha \ )^2 \cos^2\alpha \ \} \ . \tag{7}$$

If the pitch of the projected fringe angle is denoted by αo and the phase error $\Delta\phi$ is in degree, $\Delta\alpha$ is expressed as

$$\Delta\alpha = \alpha o \ \Delta\phi \ / \ 360 \ . \tag{8}$$

From these equations we can calculate ΔZo on the basis of the simulated results.

Figure 7 is an example of the calculated results which corresponds to the experimental arrangement in section 3. The condition is $\xi = 0$, $\alpha = \pi/4$, a = Zo = 672mm and N = 4. In the figure, αo is expressed in the dimensional pitch on the object surface.

The data plotted in the figure is the experimental result obtained using the arrangement with the same condition as the calculation. In the experiment, the object is a flat plate. ΔZo is the mean square value of the variation of the measured height Zo after compensating the inclination of the plate.

This result shows that in the present experiment the errors correspond to the phase error between 4bit and 5bit for large projection pitch. This result seems to show that the phase shift error is a dominant source of error.

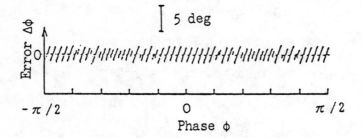

Fig.5 Phase error Δφ caused by digital
signal convertion of four-step
phase shift

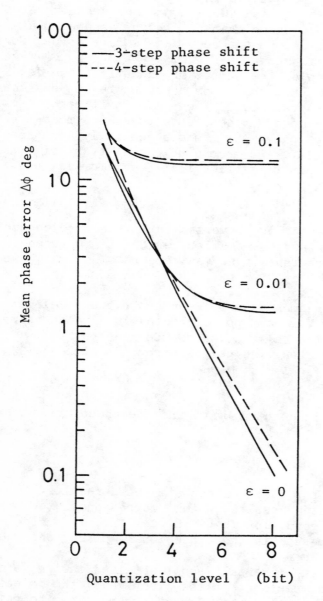

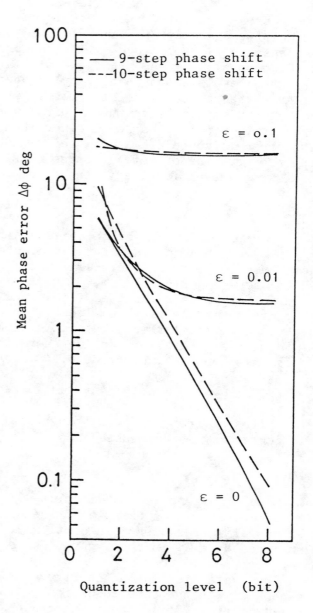

Fig.6 Dependence of phase error Δφ upon the quantization level and phase-shifting
error ε the number N of phase shift

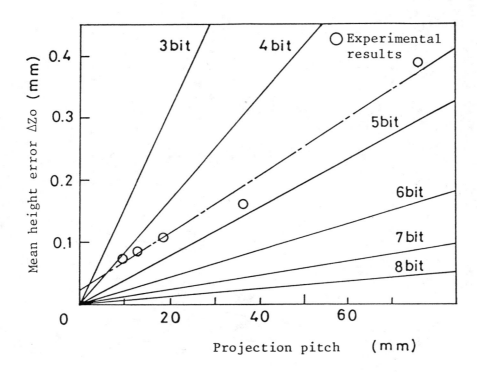

Fig.7 Mean height error ΔZo due to projection pitch
of interference fringe for the present
experimental system

4.3 Measurable spatial range

Two coherent light waves make fringes of good contrast in space as far as the two waves are superposed. Thus the measurable spatial range is not restricted by contrast degradation of the projected grating. The range is mainly determined by the depth of focus of the TV camera. It determines the spatial lateral resolution of the image.

Even if high lateral resolution is not required, large circle of confusion causes contrast degradation of the detected fringes, resulting in the degradation in height measuring accuracy. If the fringe spacing is denoted by P and the radius of the confusion circle is denoted by r, contrast degradation ratio γ of the fringe is calculated as

$$\gamma = 2 \ J_1 \ (2\pi r/P) \ / \ (2\pi r/P), \qquad (9)$$

where J_1 is the first order Bessel function. This formula is derived by assuming the fringe intensity is integrated in the interior of the confusion circle.

Using this formula and Fig.6, we can calculate the height error as a function of the depth object. Figure 8 shows a calculated result for the optical system used for the experiment.

In the same figure, experimental results are also shown. The experiments were performed by placing a plane object at the position of varying distance from the camera and mean square error is calculated. The measurement is made for the two different aperture diameters. The result shows the change of error with the focus depth can be attributed to the contrast change of the fringes.

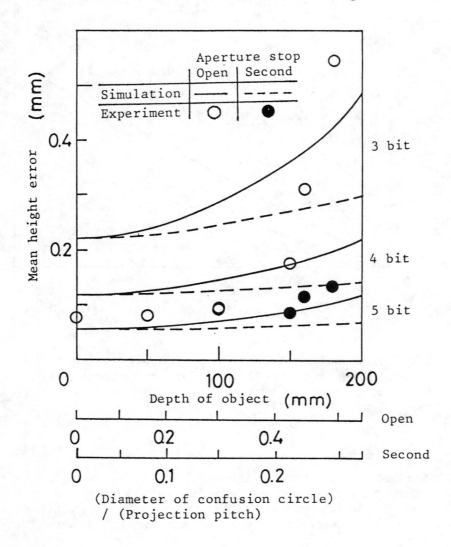

Fig.8 Mean height error due to the depth of object
(Projection pitch = 14.4 mm)

5. Conclusion

A system for profiling an object with dimensions in the order of tens of cm is constructed. It based on phase-shifted interference fringe projection method. Some simple objects are measured with sub-mm precision.

Simulations clarified the relation of the phase measuring error with quantization level in the detector and error in the phase-shifting device. On the basis of the simulated result, error in height is discussed and the result is compared with experiment, which confirms the discussion. Spatial measuring range is also discussed based on the same basis and compared with experiment.

6. References

1. M.Takeda and K.Mutoh,"Fourier Transform Profilometry for the Automatic Measurement of 3-D Object Shapes", Appl.Opt.,22, 3977-3982 (1983)

2. T.Yatagai and T.Kanou, "Aspherical Surface Testing with Shearing Interferometer Using Fringe Scanning Detection Method", Optical Engineering, 23, 357-360 (1984)

3. K.Sato and S.Inokuchi, "Three-Dimensional Surface Measurement by Space Encoding Range Imaging", Journal of Robotic Systems,2,1, 27-39 (1985)

4. T.Yoshizawa and K.Suzuki, "Automatic 3 D Measurement of Shape by Grating Projection Method", (in Japanese) Journal of the Japan Society Precision Engineering, 53, 422-426 (1987)

5. J.H.Bruning, D.R.Herriott, J.E.Gallagher, A.D.White and D.J.Brangaccio, "Digital Wavefront Measuring Optical Surfaces and Lenses", Appl.Opt., 13, 2693-2703 (1974)

6. K.Creath,"Phase-Measurement Interferometry Techniques", Progress in Optics XXVI, Ed.E.Wolf, Chapter V, Elsevier Science, USA, (1988)

7. V.Srinivasan, H.C.Liu and M.Halioua, "Automated Phase-Measuring Profilometry of 3-D diffuse Objects", Appl.Opt., 23, 3105-3108 (1984)

8. S.Kakunai, K.Iwata, M.Hasegawa and S.Yamaguchi, "Grating Projection Method for Profile Measurement Using Phase-shifting Technique", (in Japanese) Journal of the Japan Society Precision Engineering, 55, 141-145 (1989)

Analysis of solute concentration and concentration derivative distribution by means of frameshift Fourier and other algorithms applied to Rayleigh interferometric and Fresnel fringe patterns

A.J.Rowe*, S.Wynne Jones#, D.Thomas* & S.E.Harding#

National Centre for Macromolecular Hydrodynamics, *University of Leicester, Leicester LE1 7RH, U.K. and #University of Nottingham, Sutton Bonington, LE12 5RD, U.K.

ABSTRACT

The equilibrium distribution of particles dispersed in an aqueous solute situated in a centrifugal accelerative field is routinely studied by means of an optical trace recorded photographically. Rayleigh interferometric fringe patterns have been widely used to give this trace, in which the displacement of the parallel fringes is directly related to particle concentration differences. We have developed a simple but highly efficient frameshift algorithm for automatic interpretation of these patterns[1]. Results obtained from extensive use and further definition of this algorithm confirm its validity and utility.

We have also studied algorithms for the interpretation of Fresnel fringe patterns yielded by an alternative optical system. These more complex patterns involving non parallel fringes can be analysed successfully, subject to certain conditions, with a precision similar to that obtained using Rayleigh interference optics.

1.INTRODUCTION

1.1 The methodology employed

The Analytical Ultracentrifuge is used to study systems of dispersions of large particles in a fluid (usually aqueous) medium[2]. From the results yielded, important conclusions can be drawn concerning the structure, interactions and state of dispersity of the particles. A particularly powerful approach involves the balancing of radial centrifugal forces acting on the particles against the forces arising from the chemical potential gradient induced by the former. The sytem in this state is said to be at 'sedimentation equilibrium'.

The basic equation governing the distribution of particles at sedimentation equilibrium in a centrifugal field is

$$dc/dr \quad = \quad k\,c\,r \tag{1}$$

where c is the concentration of particles at radial position r, and k is a reduced particle mass, given by

$$k \quad = \quad M_r(1 - \bar{v}\rho)\,\omega^2/RT \tag{2}$$

where M_r is the particle mass, $\bar{v}$ the partial specific volume ($\approx$ reciprocal density) of the particles, ρ the density of the fluid, R is the gas constant, ω the angular velocity and T the temperature (deg Kelvin).

Equation (1) is frequently used in integrated form :

$$\Delta \log c \, / \, \Delta(r^2) \quad = \quad k \, /2 \tag{3}$$

Normally the parameter M_r will be the object of study. It may itself be a function of r, either directly as a result of polydispersity or indirectly as a result of depending upon c, which as a consequence of redistribution of the solute particles varies with r.

The basic function of any analysing optical system is thus to record a pattern capable of being interpreted to yield *either c or* dc/dr as a function or r. In the former case equation (3) is applicable : in the latter case equation (2) would be used.

1.2 Analysis to yield c vs r data

The classical approach has been to use Rayleigh interference optics to give a pattern in which the displacements of the fringes in a direction (z) normal to radial is a linear function of the concentration increment at the radial position in question. The fringes are of course equi spaced and parallel, and hence a scan across them in the z direction yields a sinusoidal intensity function whose phase is a measure of (the non integer part of) the fringe shift.

We have developed a simple but fast and stable algorithm for deriving the phase shift from the intensity function[1]. The latter is logged from the photographic record of the fringe pattern, using a commerical scanning densitometer, the LKB 2202 laser densitometer. Then if Q fringes are contained within the window analysed, an iterative frameshift is performed within the data set, to maximise the Fourier coefficient of order Q. The method is thus a null method, which searches for the frameshift which will set the phase term to zero[1].

Thus this algorithm, unlike earlier approaches in this area, yields estimates for the fringe increment whose precision is not a function of the latter. The precision of the recorded fringes may be gauged subjectively from Figure 1 :

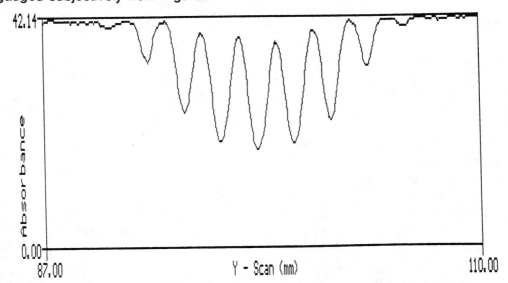

Figure 1 Digitised optical density values outputted from an LKB Ultroscan XL 2222 (two dimensional scanner) scanning at a single radial position. 575 values were logged in this case at each of 175 radial positions in the cell, and these form the data set for subsequent analysis. Data for sedimentation equilibrium experiment on lipase, using Rayleigh interference optics.

Initial results using our algorithm suggested a precision of f/500 (f is a single fringe increment) as being attainable. Further development and applications of the algorithm have followed, and are now presented and discussed.

1.3 Analysis to yield dc/dr vs r data

The earliest optical method used to analyse distributions within the ultracentrifuge cell was the 'Schlieren' optical system, in which an analysing diaphragm is inserted into the back focal plane of the camera lens employed to image the cell. Shadows or other traces are produced, whose displacement, again in the z direction, is proportional to the first derivative of solute concentration with respect to radial distance. Other than in the earliest work, a 180° phase plate has been used as analysing diaphragm. The resulting single trace is rather broad as compared to an interference fringe (Figure 2).

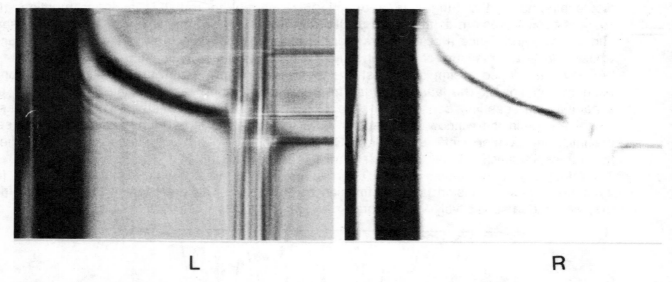

L   R

Figure 2 Phase plate Schlieren records of a solution at sedimentation equilibrium in an ultracentrifuge cell (MSE Mk II Analytical Ultracentrifuge). The solution column is some 2 mm long in real space. From original negatives conventionally (L) and correctly (R) exposed.

It has been universally considered that the precision with which this trace can be interpreted falls well short of what can be achieved using Rayleigh intereference optics. Subjectively this is understandable. The Schlieren trace appears relatively broad, and only a single trace is yielded, thus making unavailable the reduction in noise/signal normally achieved from multiple records. Yet the principal optical components of the two optical systems are identical, and are used at the same working aperture. Insofar as distinctive components are introduced in either method, there is no reason to suppose that these limit the information transfer function, which one would expect to be very similar in both cases, given adequate interpretative algorithms.
We have therefore researched the possibility of developing the interpretation of Schlieren records to a much higher level than heretofore. There are a number of practical reasons for doing this. As detailed below, we find that with suitable developments of the methodology, and subject to certain relatively minor reservations, results from the Schlieren optical method can indeed be

interpreted with a precision approaching those obtained by the Rayleigh interference method. The basis of this is the recording and interpretation of the more complex *Fresnel* fringe patterns generated by the Schlieren optical diaphragms. Several approaches to the interpretation of Fresnel fringe patterns can be defined. It seems likely that an optimal approach has yet to be delineated, but results to date are more than adequate to demonstrate the potential of work in this area.

2. FRINGE SHIFTS IN RAYLEIGH INTERFEROMETRIC PATTERNS

We have completed the construction of a 2 dimensional data acquisition system and the writing of a package of user friendly interpretative software, built around the frameshift Fourier algorithm described earlier[1]. Sophisticated search procedures have been incorporated to ensure that the system reproducibly and stably finds the correct fringe intensity maximum in what is now a full 2 dimensional record (cf our earlier version[1] which was a series of individual one dimensional scans) . As it is now possible to analyse data at up to 200 radial positions from a single experiment, rigorous tests can be performed to assess such factors as sample homogeneity and interactions (Figure 3) :

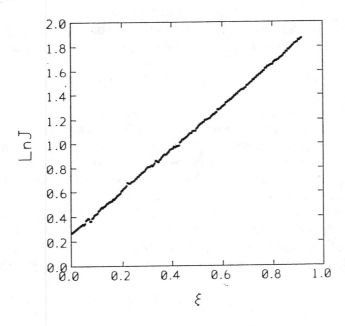

Figure 3 Plot of the logarithm of the solute concentration (expressed in absolute fringe numbers, J) versus the radial displacement (squared) parameter ξ . Data from a low speed sedimentation equilibrium experiment on recombinant Hirudin, loading concentration 0.8 mg/ml. From the slope a weight averaged molecular mass of 7080 ± 100 is computed (from sequence = 6964).

The completed system is now in intensive use, and results on many systems have fully justified our initial estimates[1] of the precision attainable.

3. REFRACTOMETRIC OPTICS AND FRESNEL FRINGE PATTERNS

3.1 Refractometric optics

The presence of an analysing diaphragm in the back focal plane of the camera lens imaging a cell, in this case located in an ultracentrifuge rotor, leads to the formation of a 'Schlieren' pattern in the image plane. The presence of a cylindrical lens results in a pattern in the image plane in which the z deflection of the trace at any radial position is linearly related to the refractive index change (dc/dr) in the conjugate locus in the cell plane (Lloyd[3]). Although any physical form of diaphragm can in principle be employed, it has long been customary to use a 180° phase plate (Wolter[4]), which causes no loss of transmitted light and has been considered on general principles (rigorous analysis appears not to have been performed) to maximise the information transfer function[3]. As a pure phase plate records no signal as dc/dr > 0, a thin line of metal evaporated onto the half wave step is normally added[3]. The latter *only* produces a trace for zero or very low dc/dr values. The optics of the transition region of dc/dr have yet to be been defined.

The resulting trace shows a well defined but rather broad line (Figure 1). This represent the zeroth order fringe of an often poorly resolved Fresnel pattern. We have addressed ourselves to the definition at high resolution of the co ordinates of this pattern, both by location of the zeroth order fringe, and by an alternative approach in which we derive and apply a relationship between Fresnel fringe *spacing* at defined r and the corresponding dc/dr and Δc values.

3.2 Definition of the zeroth order fringe in a Fresnel pattern

As noted above, the conventional approach here has been to record the Schlieren pattern using a modified phase plate diaphragm. The pattern is in principle symmetric with respect to the zeroth order fringe, which is located in the centre of the line trace. However, as was noted by Rowe and Khan[5], a simple knife edge diaphragm has a better established optical theory, and can yield results of a precision equal to that given by a phase plate. We have therefore explored the use of a simple knife edge diaphragm to generate the Fresnel pattern.

We have in all cases adopted certain modified procedures for setting up of the ultracentrifuge cells and optics designed to maximise the information transfer function. These will be described elsewhere. They are not relevant to the interpretative algorithms described below, but are indispensible if results of the precision described are to be obtained in practice.

3.2.1 The zeroth order in a phase plate generated Fresnel pattern

The zeroth order in this case is simply the maximum density (minimal intensity) in the z scan across the line pattern (Trautman and Burns[6]). Unfortunately as normally recorded the line is rather broad and the precision in the estimation of the position of the maximum not high. This is the conventional reason behind the general opinion that Schlieren optics are inherently of low precision.

Several experimental procedures can be adopted which largely circumvent the problem of the width and lack of definition of the optical trace. The experimental procedure most relevant to the present discussion is the use of long photographic exposures to bring the optical record *near to the centre of the line trace* into the linear part of the gamma curve of the recording material. It seems not to have been generally appreciated (though Lloyd[3] commented briefly on the matter) that giving a 'normal' exposure as gauged for the whole photograph seriously degrades the transfer function in the critical region.

Locating the zeroth order in a suitably exposed photographic image (i.e. highly *over* exposed with respect to 'non information') by analysis of successive radial z scans results in a very smooth data set. Figure 3abelow illustrates typical final results computed from data measured in this way.

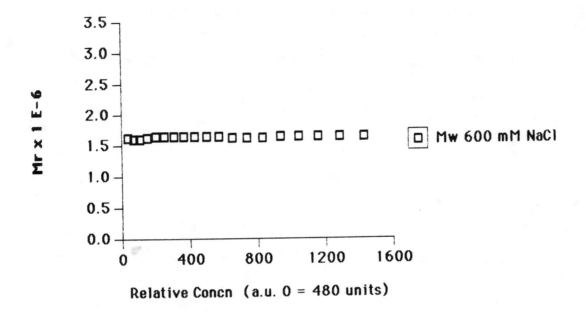

Figure 3a $M_{w,r}$ values computed for the protein dynein at a cell loading concentration of 0.5 mg/ml. These point weight average values are computed by numerical integration of the dc/dr values to give c values, the constant of integration being found by a numerical manipulation based upon the equivalence of harmonically related averages.

These results are entirely comparable to those which would be obtained using Rayleigh interferometric optics together with the Fourier algorithms described above, and would correspond to a precision of at least f/300 in the latter terms. There are however certain qualification which must be made concerning the absolute *accuracy* as distinct from the *precision* of the results. As noted above, for low values of dc/dr, the optical behaviour of a compound phase plate is far from well defined. This can be circumvented by avoidance of such conditions. More seriously, it seems not to have been appreciated that the physical properties of the phase plate and in particular its phase angle are critical.

We have computed intensity distributions in the image plane for a phase plate of various angles, using the Cornu Spiral construction[6]. The treatment given by Trautman and Burns[6] assume a phase angle of 180⁰. We have extended this to the more general case, and as the tabulated values for the Cornu Spiral are in some cases of insufficient resolution we have computed the co‑ordinates from the relation[7] :

$$F(x) = (2/\pi)^{0.5} (x + jx^3/3 + (j^2/2!)(x^5/5) + \ldots + (j^n/n!)(x^{2n+1}/2n+1))^{0.5} \quad (4)$$

for $n = 0,1,2 \ldots$ and the spiral is plotted in the complex plane $j = (1)^{0.5}$

The results are shown in Figure 4 below for phase angles of 180, 150 and 120 degrees :

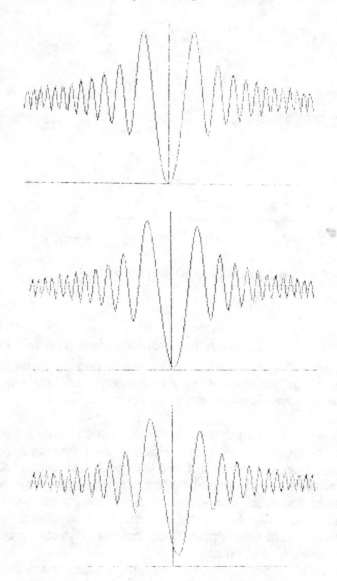

Figure 4. Computed intensity distribution of the Fresnel pattern yielded by a phase plate of phase angle 180 degrees, 150 degrees and 120 degrees. The location of the true geometrical edge is shown by the vertical line in each case.

It is clear that the location of the true geometrical edge coincides with the minimum of the intensity distribution *only* for the case of a 180 degree phase angle. Moreover the pattern is not symmetrical for other phase angles. Thus the wavelength used is critical, and must be tuned to the particular phase plate used, a precaution which has not formed part of normal practice, and can give rise to practical problems in securing adequate light intensity in monochromated light.

3.2.2. The zeroth order from a knife edge pattern

A simple knife edge used as a diaphragm results in a 'shadow' pattern, with a set of Fresnel fringes (Figure 5). Although actually of slightly smaller amplitude than those generated by a phase plate[6], th lower dynamic range of the image means they are frequently better registered.

Figure 5. A part of a Schlieren pattern recorded with a knife edge diaphragm from a solution of an enzyme (chloramphenical acetyl transferase) at sedimentation equilibrium. Centrifugal direction is from left to right.

The location of the zeroth order fringe can be computed by measurement of (say) the 1st and 2nd order minima, and from the knowledge that their location (z) with respect to the zeroth order is given for order i by

$$z = (4(i + 1) - 0.5)^{0.5} \qquad (5)$$

In practice only the first two or three orders can be measured in a z scan (Figure 6) :

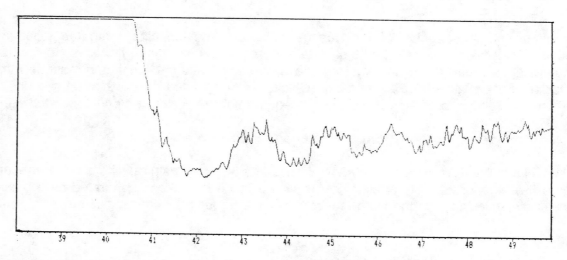

Figure 6 Vertical (z) scan across a set of fringes from a knife edge diaphragm (as in Figure 5) The location of the centre of the peaks in the scans can be determined by established procedures[1]. The resulting precision is found to be of the order of 1 to 2 % of the fringe spacing.

3.3 Zeroth order determined from the Fresnel fringe spacing

It is possible to use the values of the fringe spacings in the z direction to evaluate Δc *directly* instead of by computation of the location of the zeroth order. This is because of the defined relationship[8] between fringe spacing and the second derivative of the refraction (and hence concentration) gradient, from which it follows at once that

$$\Delta c = \int_{r_a}^{r_b} \int \{ (\sum_{i=1}^{i=(n-1)} (z_{i+1} - z_i) / (4(i+1) - 0.5)^{0.5} / (4i - 0.5)^{0.5})) / (n-1) \} \quad (6)$$

This equation defines the relationship between Fresnel fringe spacings and Rayleigh fringe shifts. The double integration is highly favourable with respect to error reduction in the data set, achieving at least an order of magnitude of diminution, greater if the summation can be effected by measurement of multiple fringes. A constant of integration is required for the first integration. This is in fact the zeroth order spacing, but as an independent estimate of this constant can be obtained from each radial scan, errors in its estimation are not serious.

We have evaluated several procedures for determining the z spacings in a multiple fringe pattern. Direct fitting of the Fresnel function by a least squares algorithm has been implemented, but is not totally successful. This is primarily because the intensities in a Fresnel pattern, ranging from true zero upwards, cannot possibly be recorded photographically within the linear part of the gamma curve of the emulsion. The true pattern is thus convoluted with an envelope function, which as the Fresnel amplitude/intensity function is anharmonic, cannot readily be deconvoluted as with Rayleigh patterns[1]. Furthermore, as noted above, although up to 10 or more Fresnel fringes can be discerned by eye, the intensity scans give only 2 or 3 clearly defined maxima/minima (Figures 5 & 6).

Thus this approach, whilst somewhat superior to the simple evaluation of the zeroth order, has yet to be developed to its full potential.

4. CONCLUSIONS

The evaluation of relative solute concentrations within an ultracentrifuge cell by Rayleigh interferometric fringe shifts can, using optimal procedures and interpretative algorithms, yield a data set with a precision approaching f/500, where f is a single fringe shift[1]. Our results to date using the alternative Schlieren optical system show that under identical experimental conditions (i.e. same solute concentration and optical path length) this latter system can approach the Rayleigh fringe level of precision. In terms of Rayleigh fringe shifts, the direct evalution of the zeroth order by a phase plate diaphragm (3.2.1) attains f/300, though with some danger of systematic error, and by knife edge diaphragm attains f/150 to f/200. By the use of the fringe spacings (3.2.2) and the transformation noted above (equation 6) a precision of close on f/300 is attained.

Refractometric (Schlieren) optics have a number of advantages over Rayleigh interferometric optics. Their alignment is much simpler, and window distortions are a much less serious problem. The widely held supposition that their precision is much inferior to Rayleigh optics lacks a theoretical basis and is now shown to be untrue in practice. Given more advanced two dimensional analysis of the recorded fringe patterns to enable up to 10 fringe spacings to be analysed, the precision of refraction and hence concentration increment determinations will be essentially the same by either method, and it will be possible to choose the simpler refractometric system when experimental conditons so dictate. It is possible that the transformation between the two types of fringe pattern noted above and the interpretative algorithms developed may have application to other systems.

ACKNOWLEDGEMENTS

We are indebted to the Science and Engineering Research Council (U.K.) for support of this work.

REFERENCES

1. S.E.Harding and A.J.Rowe, "Automatic Data Capture and Analysis of Rayleigh Interference Fringe Displacements in Analytical Ultracentrifugation", Optics & Lasers in Engineering, vol. 8, pp. 83-96, 1988.

2. T.Bowen and A.J.Rowe, <u>An Introduction to Ultracentrifugation</u>, Wiley, London, 1970.

3. P.H.Lloyd, <u>Optical Methods in Utracentrifugation, Electrophoresis and Diffusion</u>, Oxford, 1974.

4. Von H. Wolter, "Verbesserung der abbilden Schlierenverfahren durch Minimumstrahlkennzeichnung", Ann. der Physik, vol. VI.7, 182-192, 1950.

5. A.J.Rowe and G.M.Khan, "A Comparison of Results from Phase Plate and Knife Edge Diaphragms in Conjunction with the Schlieren Optical System in an MSE Analytical Ultracentrifuge", Rev. Sci. Instr. vol. 42, pp. 1472-1474, 1971.

6. R.Trautman and V.W.Burns, "Theory and Test of Commercially Available Wolter

Phaseplate for Use in Schlieren Optical Systems Employed in Ultracentrifugation and Electrophoresis", Biochem. Biophys. Acta, vol. 14, pp. 26-35, 1954.

7. A.Papoulis, <u>Systems and Transform with Application to Optics</u>, pp. 70-73, McGraw Hill, Maidenhead, England, 1968.

Managing some of the problems of Fourier Fringe Analysis

D.R. Burton and M.J. Lalor

Coherent and Electro-Optics Research Group,
Liverpool Polytechnic, U.K.

ABSTRACT

Recent work has shown the importance of the fast Fourier transform (FFT) in the automatic analysis of fringe patterns. Three problem areas are encountered when the FFT is used as an analysis tool in this fashion, namely, aliasing, the picket fence effect and leakage. The authors consider leakage to be by far the biggest problem in this application. The paper defines what causes leakage and suggests how leakage may be controlled using data weighting functions. Comparative results obtained using a variety of functions are presented.

2. INTRODUCTION

Recent work by the authors and others has shown the importance of the fast Fourier transform (FFT) in the automatic analysis of fringe patterns[1-3]. This has led to the development of a technique for non-contact precision measurement of surfaces which has a wide variety of applications[4,5].

This paper reviews the Fourier fringe analysis technique for the reconstruction of surfaces underlying optically generated contour maps. The technique views the map as a constant spacing straight fringe pattern phase modulated by the underlying surface shape. The Fourier transform is used as a means of demodulating the pattern, producing a 2π wrapped phase distribution. This phase information is then converted into a range map which is amenable to sophisticated analysis, using techniques such as differential geometry, to arrive at measurements of surface parameters of functional importance to engineers.

The FFT, which is an efficient method to compute the discrete Fourier transform (DFT), contains a hornet's nest of traps for the unwary user. In fact there are three major sources of error associated with the use of the FFT, namely, aliasing, the picket fence effect and leakage. Mostly engineers use the FFT blindly to approximate the continuous Fourier transfrm (CFT). Using the transform in this manner can lead to problems which are caused by a misunderstanding of what the approximation involves. This paper describes all three sources of error, but concentrates on the error effects of the leakage phenomenon and attempts to control these leakage effects using various data weighting functions.

3. A REVIEW OF THE TECHNIQUES

It can be shown that for a situation, such as in Fig. 1, the intensity profile of the illuminating fringes as seen by the camera is given by:

$$I(i,j) = \frac{I_{max}}{2} \cos \left[\frac{2\pi \cos \theta}{p_o} j + \frac{2\pi \sin \theta}{p_o} h(i,j) \right] + \frac{I_{max}}{2} \tag{1}$$

which can be seen to be a phase modulated cosine function of carrier frequency $\cos \theta / p_o$ and modulation $h(i,j)$.

Taking the Fourier transform of equation (1) would produce a power spectrum such as that shown in Fig. 2. Several points need to be made regarding this spectrum:

(i)   The large central zero order peak represents the average brightness of the image. It also absorbs all of the low frequency noise effects such as slowly varying image intensity etc.

(ii)   The two symmetrical first order peaks are related to our sinusoidal function equation (1). They are not impulse functions, but have a spread which is due to the modulation term h(i,j).

(iii)   Beyond these peaks lies only higher frequency noise due to speckle etc.

Filtering the Fourier domain in such a fashion as to remove the zero order peak and one of the first order peaks yields a non-symmetrical spectrum which on inverse Fourier transformation will produce non-zero real and imaginary parts[6].

The phase of the original fringe pattern may be calculated from this real/imaginary data using the expression:

$$\phi(i,j) = \arctan \frac{I_m(i,j)}{R_e(i,j)} \tag{2}$$

Equation (2) gives the phase distribution modulo 2π. This must be unwrapped to create a continuous phase distribution from which a range map can be evaluated using an inversion of equation (1)[5].

The resulting range map can be analysed using a variety of techniques depending upon the surface parameters required[7-9]. The method currently employed by the authors involves convolving the range map with suitable differentiating filters to produce first and second order derivatives. These derivatives may then be used in certain expressions from the theory of differential geometry to calculate the first and second fundamental forms of a surface[10]. Knowledge of these parameters allows calculation of any geometric property of the surface at any point.

4. DEFINITION OF PROBLEM AREAS

As mentioned previously there are three major sources of error in the use of the FFT. These are defined below.

Aliasing is the phenomenon due to which high frequency components of a time or spatial function can translate into low frequencies if the sampling frequency is too low. The problem can be alleviated by increasing the sampling frequency, but aliasing may still be present even if the highest frequency component is less than half the sapling frequency.

The picket fence effect occurs if the analysed waveform includes a frequency which is not one of the discrete frequencies (an integer times the fundamental). A frequency lying between the nth and $(n+1)^{th}$ harmonics affects primarily the magnitude of the nth and $(n+1)^{th}$ harmonics and secondarily the magnitude of all other harmonics.

Leakage refers to the apparent spreading of energy from one frequency into adjacent ones. It arises due to the truncation of the time sequence such that a fraction of a cycle exists in the waveform that this subjected to the FFt. This is different from the CFT, where just by virtue of the fact that the waveform is truncated, whether it is a fraction of a cycle or not, leakage will occur. A true CFT is taken over the range $\pm\infty$. The DFT, and hence also the FFT, must truncate this sample to one finite length. Implicit in this truncation is the assumption that the finite sample contains the fundamental frequency of a long wavelength periodic signal ile. the sample is assumed to repeat infinitely in both the positive and negative directions. Thus any mismatch between the ends of the sample - caused by having a fraction of a cycle - will distort the result of the FFt from the idealised CFT. Thus in the frequency domain the actual transform obained is the convolution of two transforms - those of the windowing function (usually rectangular) and the waveform to be analysed.

5. SOLVING THE LEAKAGE PROBLEM

Repeating the statements of the previous sections:

1. Leakage is caused by the sample being analysed not being periodic.
2. Data are always windowed. The basic form is the rectangular window of constant unity amplitude of duration N, where N is usually a power of two.

Given these facts, three methods have been proposed for coping with the leakage problem.

(i) Choose a window larger than the data sample to be analysed and "pad" the redundant capacity with zeros.

(ii) Choose a window larger than the data sample to be analysed and "pad" the redundant capacity with synthetic data interpolated from the data under analysis. The interpolates fill the data window with exactly an integer number of waveform cycles.

These two methods are not applicable in many cases, especially where the data are almost periodic or where the frequency domain data are to be further proceesed (as i the case of Fourier fringe analysis). They can, however, be applied to make the data at least begin and end at the same level. This prevents jump discontinuities at the window edges. Eliminating these discontinuities will not entirely eliminate leakage, but will certainly help to reduce it.

The third method is again designed to reduce discontinuities at the window edges by tapering the rectangular window, i.e. removing the constant unity amplitude. The object is to remove the abrupt edges of the rectangular window and allow them to fall off smoothly to zero. This is achieved by multiplying the acquired data with a data weighting function.

6. DATA WEIGHTING FUNCTIONS

Data weighting functions, or windows to give them their more common title, are basically a technique for minimising the effects of time domain truncation. There are a number of popular windows available, all aimed at reducing leakage by reducing the amplitude of the first order peaks in the frequency domain of the window itself.[11,12]

For the purposes of this investigation five of the more commonly encountered windows have been used. These are:

1. Rectangular
2. Extended Cosine Bell
3. Hanning

4. Hamming
5. Cosine^4.

These are shown in one dimensional form in Fig. 3.

These, and most other windows generally encountered, were originally designed to be used in one dimensional temporal analysis. However, it has been shown that good two dimensional spatial symmetrical windows can be obtained from one dimensional windows using the relationship:

$$w'(x,y) = w[\sqrt{x^2 + y^2}] \qquad \text{if} \qquad \sqrt{|x^2 + y^2|} \quad < \quad \frac{T'}{2} \tag{3}$$
$$= 0 \qquad\qquad\qquad\qquad \text{otherwise}$$

where $w(.)$ is centred at $[x=0, y=0]$ and T' is the truncation interval.[13]

The windows are evaluated over a 64 × 64 array space and applied by multiplication with incoming 64 × 64 pixel array data selected under software control from a 512 × 512 image.

7. EXPERIMENTAL OBSERVATIONS

For comparison purposes a simulated sphere of radius 5000 pixel units was generated to provide the surface data. These data were analysed using each of the data weighting functions and a surface reconstruction and mean radius evaluation obtained for each case.

Figs. 4 and 8 show horizontal and vertical centre line profiles superimposed on expected profiles and an error curve for each case.

Quantitative results for mean principal radii over the whole 64 × 64 pixel array (4096 points) are given in table 1.

Window	Rectangle	Bell	Hamming	Hanning	Cosine^4
Nominal radius	5000.000	5000.000	5000.000	5000.000	5000.000
Horizontal mean radius	4973.649	5004.351	5002.895	4998.373	4999.473
Vertical mean radius	4973.250	5003.955	5002.498	4997.976	4999.076
Horizontal mean error	-0.527%	0.087%	0.058%	-0.033%	-0.011%
Vertical mean error	-0.535%	0.079%	0.050%	-0.041%	-0.018%

Table 1

8. INTERPRETATION OF RESULTS

Firstly to clarify the analysis undertaken and to explain why the results are presented in the form of radii of a sphere. Given the ultimate aim - a technique for the precision determination of engineering surfaces - the dimensional measurement of spherical radius provides a benchmark test for three reasons:

(i) A sphere possesses constant directionality. Although the sphere displays curvature in two principal directions this curvature (and therefore) radius is constant at every point, i.e. there is no directional favourability.

(ii) Curvature of a sphere is represented by second order derivatives. Basic mathematics show that differentiation has the effect of attenuating or magnifying any errors present. Proving the technique using a second order differential function demonstrates robustness.

(iii) The sphere is a commonly used and well understood function. Errors tend to be immediately obvious to the observer.

Results obtained using all the data weighting functions, both those presented here and others investigated, are broadly acceptable with the cosine^4 function providing a mean minimum error. This can be seen in Figs. 4(a) to 8(a).

The very small differences between horizontal and vertical mean error can be discounted and, as expected, show no signs of directional favouritism.

Of more interest is the sign of the errors. The Extended Cosine, Bell and Hamming windows produce positive distortion of the expected result whereas the basic rectangle, Hanning and cosine^4 have the effect of producing negative distortion.

In Figs. 4(b) to 8(b) horizontal and vertical error curves are presented showing absolute error for panel centre lines. All cases display a complex error distribution with maximum errors at panel edges and minimum errors at the panel centre. Between edges and centre the error curves tends to oscillate as a result of the relationship between the function of the sphere and the differential fitting function.

9. CONCLUSION

Previous work has suggested that the use of data weighting functions in certain situations can cause spectral smearing.[11] No evidence to this effect has been found. The results obtained imply that the use of windows does not destroy information used in precision measurement which is contained in the Fourier domain.

Some windows produce better results than others. In fact the best window, cosine^4, produces a factor of three reduction in the error in spherical radius evaluation.

The range of windows presented exhibit minor improvements over the basic rectangular window at the cost of increased computational expense. For use in a precision measurement application this cost is acceptable. However, in a simple surface reconstruction system it probably is not.

In essence the use of windows as a data improvement technique must be justified by the individual application.

10. REFERENCES

1. T.M. Kreis, "Fourier-Transform Evaluation of Holographic Interference Patterns", Proc. SPIE, vol. 814, 1988.
2. M. Takeda, H. Ina and S. Kobayashi, "Fourier-Transform Method of Fringe Pattern Analysis for Computer Based Topography and Interferometry", J. Opt. Soc. Am., vol. 72, no. 1, 1982.
3. M. Takeda and K. Mutoh, "Fourier Transform Profilometry for the Automatic Measurement of 3D Object Shapes", Applied Optics, vol. 22, no. 24, 1983.
4. D.R. Burton and M.J. Lalor, "The Precision Measurement of Engineering Form by Computer Analysis of Optically Generated Contours", Proc. SPIE, vol. 1010, 1988.
5. G.R. Halsall, D.R. Burton, M.J. Lalor and C.J. Hobson, "A Novel Real-Time Opto-Electronic Profilometer Using FFT Processing", IEEE ICASSP, 1989.
6. R.C. Ramirez, "The FFT - Fudnamentals and Concepts", Prentice Hall, 1985.
7. D.R. Burton and M.J. Lalor, "Software Techniques for the Analysis of Contour Maps of Manufactured Components", Proc.SPIE, vol. 952, 1988.
8. P.J. Besl and R.C. Jain, "Invariant Surface Characteristics for 3D Object Recognition in Range Images", Computer Vision, Graphics and Image Processing, 33, 1986.
9. T. Kasvand, "Surface Curvature in 3D Range Images", Intl. Conf. Pattern Recognition, 1986.
10. M. Lipschutz, "Differential Geometry", Schaum's Outline Series, McGraw-Hill, 1969.
11. E.O. Brigham, "The Fast Fourier Transform and Its Applications", Prentice Hall, 1988.
12. A.H. Nuttall, "Some Windows with Very Good Sidelobe Behaviour", IEEE Trans. Acoustics, Speech and Signal Processing, vol. ASSP29, no. 1, 1981.
13. T.S. Huang, "Two-dimensional Windows", IEEE Trans. Audio and Electroacoustics, vol. AU-20, no. 1, 1972.

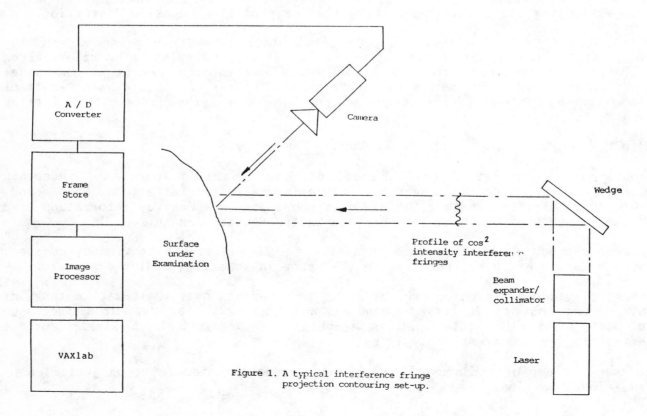

Figure 1. A typical interference fringe projection contouring set-up.

Fourier Power Spectrum

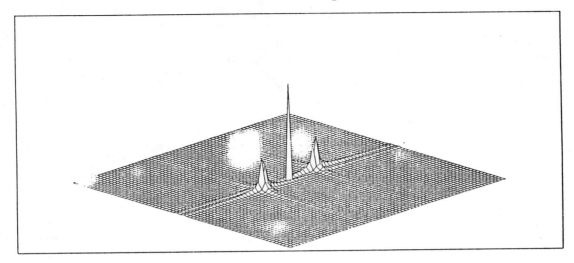

Figure 2.

Window Profiles

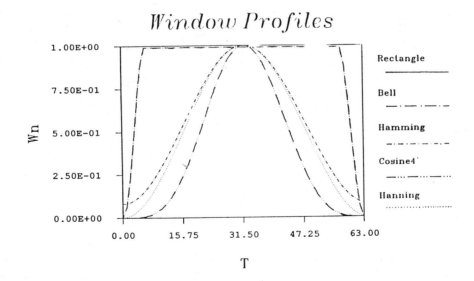

Figure 3.

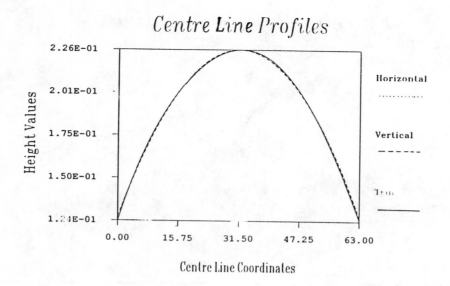

Figure 4(a). Rectangular Window

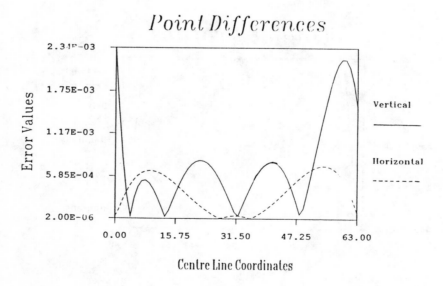

Figure 4(b). Rectangular Window

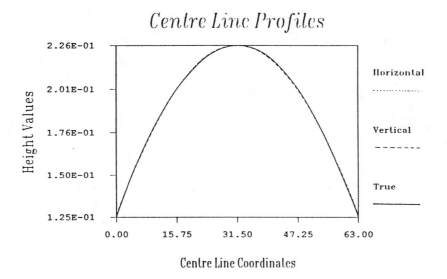

Figure 5(a). Extended Cosine Bell

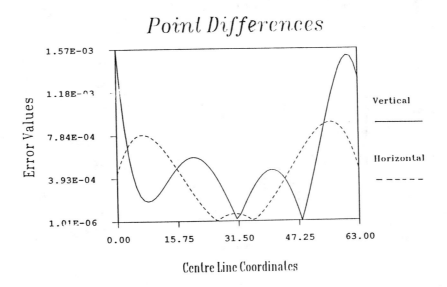

Figure 5(b). Extended Cosine Bell

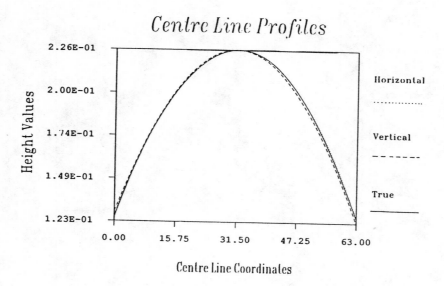

Figure 6(a). Hamming Window

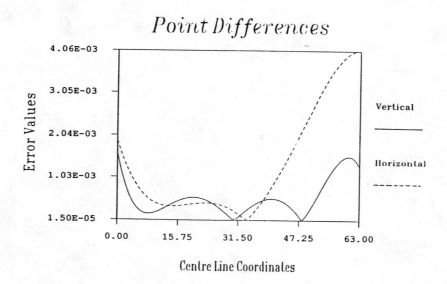

Figure 6(b). Hamming Window

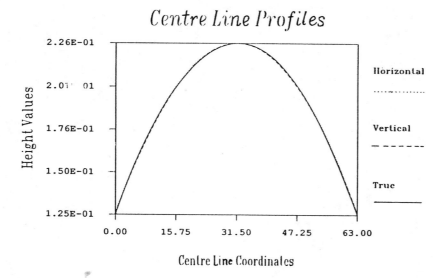

Figure 7(a). Hanning Window

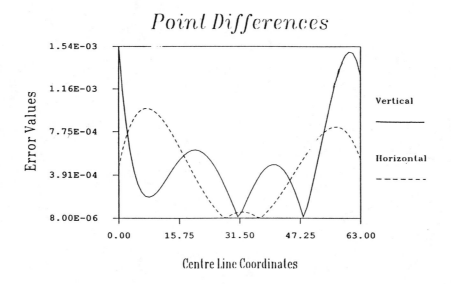

Figure 7(b). Hanning Window

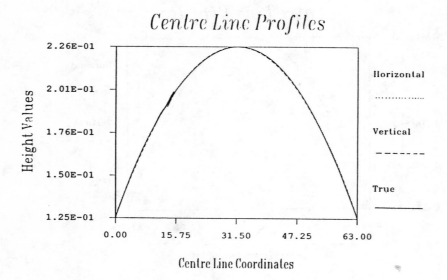

Figure 8(a). Cosine4 Window.

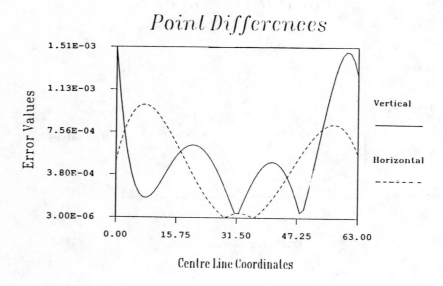

Figure 8(b). Cosine4 Window

Multiwaveband Phase Interferometer

B.R. Hill,

Group Research, Pilkington Technology Centre, Hall Lane, Lathom,
Ormskirk England

ABSTRACT

The objectives and reasoning behind the specification of a new test instrument are analysed with brief reference to the instrument it was to replace. The theory, development and operation of a 150 mm diameter scanning multi-waveband phase interferometer is described. This entails a description of how the method of phase modulation evolved with reference to its linearity and ability to operate over a very wide range of wavelengths (.6 um - 10.6 um). The method of phase measurement and its resistance to vibration is described with a brief outline of the electronic building blocks required for the operation and calibration of the instrument. The requirements and choice of computer are discussed. It will then be shown how the above elements can be put together to form a very compact and versatile instrument. The functions of the associated software are also described both for data capture and display, which is real time, and also for analysis of the captured data. The analysis includes MTF, PSF, LSF, encircled energy, polynomial fitting, homogeneity measurement and the manipulation of stored interferograms. Results from the final instrument will also be presented.

1. INTRODUCTION

The principle IR test instrument at Pilkington is an automatic IR interferometer called WATIRL (Wavefront Analysis for Testing Infrared Lenses)[1] It was a joint development between Pilkington and the British Ministry Of Defence[1] WATIRL is an extremely powerful test instrument and is still used on a routine basis for test and development.

WATIRL, though a powerful quality control and development tool has a number of disadvantages, it is large and could with poor lenses be difficult to use. Data capture can sometimes fail with poorer lenses and it takes five minutes of computing to find out if the capture is good or not. We required an instrument that was easier to use and more flexible in the complexity of interferogram that could be analysed.

WATIRL scans the intensity pattern of an interferogram over a single point detector which is fed into a computer by means of an A to D converter. The computer is used to fit the fringe maxima and minima to the intensity distribution. WATIRL can only deal with simple straight line fringe patterns and tilt fringes have to be introduced to form a straight line fringe pattern. The analysis of the intensity information is a difficult task and many safety tests are incorporated into the software to make sure that erroneous data is not produced. To extend this software to deal with more complex patterns would be difficult and the amount of computation required would need a very fast computer to give answers in a reasonable time.

With these problems in mind a project was started to look at a replacement for WATIRL. The first part of the project was to identify the best form of interferometer to use. Since one of the objectives of the work was to make the instrument more environmentaly stable a number of common path interferometers were investigated.

The decision was made to use a phase interferometer but a new form of phase interferometer had to be found if some of the original objectives were to be met. Conventional phase interferometers use piezo-electric transducers to move the reference mirror in steps of a fraction of a wavelength (usually four steps). The piezo-electric transducers have insufficient dynamic range to cope with a twenty fold change in wavelength ie visible to far IR.

2. OBJECTIVES

The following targets were set for the new instrument:-

    (a) Capable of dealing with complex fringe patterns.
    (b) Able to work in a production environment.
    (c) To be compact with the aim of being desk top.
    (d) To be multiwaveband ie. visible and IR.
    (e) To be quick and easy to use.

3. SHEARING INTERFEROMETERS

One of the aims of the new instrument was to be environmentaly stable. Since shearing interferometers are common path instruments they would fulfil this function. The most common specification for a lens system IR or visible is in terms of its MTF (Modulation Transfer Function). The MTF can be determined from an interferogram by shearing the interferogram against itself and summing the sines and cosines of the phase differences between the sheared interferograms. The shearing interferometer produces the sheared information directly and so may give a quick way of computing MTF. For these reasons this type of interferometer was investigated.

To carry out all of the diagnostic tests required we needed an interferometer that would produce variable lateral and tilt shear. This reduced the number of options considerably as most interferometers produce only one type of shear or are restricted to a fixed amount of shear.

Considering all the aspects required our investigation left us with only one suitable interferometer called a Drew Interferometer[2] (see fig. 1). The Drew interferometer was built and tested and fringe patterns captured using the WATIRL system. However, the fringe patterns produced on the instrument would be at least as difficult to analyse as the WATIRL fringes and no solution to this problem was found. Shearing interferograms give information on the slope of the wavefront and are more difficult for visual interpretation. Immunity to vibration was still good but the larger the shear the less common the path and the less the immunity to vibration. For these reasons the shearing interferometer was abandoned and an alternative sought.

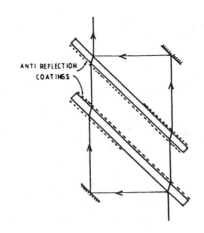

Fig. 1.   Drew Interferometer

4. CONVENTIONAL PHASE MEASURING INTERFEROMETERS

There are a number of papers on phase measuring interferometers. In this type of interferometer the optical phase is measured directly for each pixel in the interferogram. This has the advantage that any shape of fringe pattern can be analysed provided that it can be resolved by the detectors and that the slope of the wavefront is not too steep to give rise to ambiguities in fringe order. The method employed in the majority of phase interferometers that we are aware of is shown in fig. 2. A reference mirror is moved by piezo-electric transducer through an accurately known fraction of a wavelength and a measurement of intensity taken at each pixel in the detector (nearly all use CCD cameras). The reference flat is moved a further fraction of a wavelength and

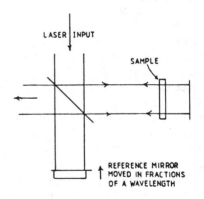

Fig. 2.   Phase Interferometer

and the intensities recorded again. This process is repeated a further twice making four intensity measurements for each pixel. For each pixel the phase is computed from the four intensities. This method requires a considerable amount of computing power to produce an answer which usually takes about 15 to 30 seconds to compute after the data is captured. There are two main methods employed, in one the piezo is moved continuously and the intensity is measured at fixed points in the reference mirrors excursion. In the second the flat is moved in steps to the required fractions of a wavelength the flat given time to settle so that there is no ringing and then the data is captured.

Neither of these similar approaches fulfilled our total objectives. They could indeed deal with complex fringe patterns, one of our major goals, but they are not as immune to vibration as we would like and are only single wave band. For this reason we set out to find an alternative method of measuring optical phase.

5. THEORY OF THE PHASE MODULATION TECHNIQUE.

Fig 3 shows a Twyman-Green interferometer into which a rotating plate has been introduced into the reference arm. As the angle of the plate is increased then so is the path length of the reference arm. The relationship between optical thickness and angle is given by:-

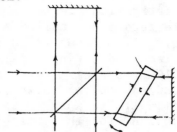

Fig. 3. Interferometer with a Rotating Plate

$$OT = 2T(NCOS(SIN^{-1}(\frac{SIN\ \theta}{N}) - COS\theta)$$

where:-
T=thickness
N=phase plate refractive index
θ=angle of rotation

This function is non-linear as can be seen from the graph in fig 4. With a single phase plate the only way to reduce the non-linearity is to increase the optical thickness of the plate. With a thicker plate a smaller angular rotation is required for a given phase shift. The phase plate was to be driven by a galvanometer drive and the data capture rate would be dependent on the speed at which the plate could be oscillated in a controlled way. Very thick plates were impractical because of their high moment of inertia.

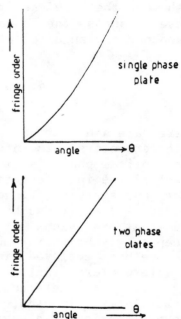

Fig. 4. Linearity of 1 and 2 plates

Since interferometers display differences in optical path then increasing the optical path in one arm is the same as reducing it in the other. By inserting a second phase plate in the other arm and decreasing its angle whilst the other plate is increasing angle then double the effect is achieved for the same rotation and also the non-linearities tend to cancel out. This method of phase shifting depending on the design of the phase plate can achieve a linearity , in the visible, of one part in 10^5 with phase shifts of over one hundred wavelengths (see fig 4).

Looking at the schematic of the interferometer in fig 5 it can be seen that the interference pattern is split into two parts by beamsplitter B2. One part of the pattern is scanned across a single point detector and the other part is reflected directly onto a single point detector such that that detector "sees" one part of the interferogram continuously. So we have one detector looking at each part of the interferogram sequentially in a TV type raster of 100 lines by 100 pixels and the other detector "looking" at one part of the pattern continuously.

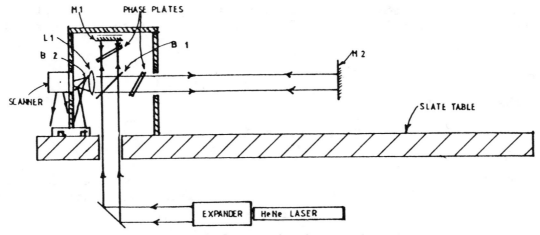

Fig. 5. Schematic of a Single Waveband Interferometer

The scan is produced by two galvanometer driven mirrors that have in built position transducers producing a feedback signal to servo amplifiers. The phase plate is driven by a similar galvanometer motor and this and the linescan motor are scanned by a common sawtooth signal. The frame is scanned by a staircase signal stepped at the end of each linescan see fig 6.

LINE SCAN

FRAME SCAN

Fig. 6. Line and Frame Scan Signals

The phase plate amplitude is such that it generates at least 100 cycles of phase shift during one line scan. If we consider an interferogram that is fluffed out and on a bright fringe fig 7 then as the phase plate is rotated the intensity of the interferogram will oscillate such that it will go:-

Bright, Dark, Bright

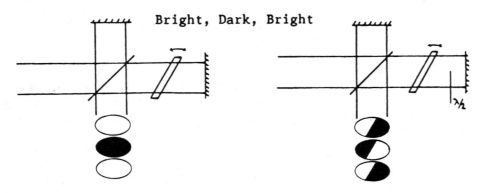

Fig. 7. Effect of a Rotating Plate on Interferograms

If we now introduce an object half way across the aperture of the interferometer such that it produces one half wave change in optical path then one half of the interferogram will become:-

Bright, Dark, Bright

whilst the other will become:-

Dark, Bright, Dark

If we now consider the two detectors, if the one that "sees" the same part of the interferogram all the time is on the bright half of the pattern it will continually receive a signal:-

Bright, Dark, Bright

The other scanned detector starts by "seeing":-

Bright, Dark, Bright

but as it crosses into the other half it "sees":-

Dark ,Bright, Dark

The electronic phase difference produced by the two detectors is the same as the optical phase difference between the two parts of the interferogram that they are looking at. This is true for all phase differences and not just the specific case of 180°, used above, to illustrate the effect.

By timing the difference between the zero crossing points of the reference and sample signals and feeding these times into a computer an aberration map of the interferogram can be built up real time. This system of phase measurement has a number of advantages over conventional techniques.

    (1) The phase plate amplitude can easily be matched to different wavelengths of light by adjusting the gain of the amplifier controlling the phase plate oscillation to obtain the correct frequency of phase shifting (in our case 8KHz).

    (2) The fact that the phase of the interferogram at one point is compared with the phase at a fixed point in the same interferogram means that the system has very good immunity to vibration. This was again one of the original objectives and has proven in practice as well as in theory.

The design of the phase plate is critical to the performance of the instrument. The moment of inertia of the plate needs to be kept to a minimum to ensure good control whilst the angle of rotation needs to be kept low to ensure good linearity.

This is particularly true in the IR. The two requirements are of course conflicting as low angles of rotation are achieved by using thicker plates. A thicker plate increases the moment of inertia for two reasons one purely because it is thicker and two because the beam goes through at an angle and if it is thicker it needs to be longer not to obscure the beam as it rotates. A computer program was developed to optimise the size of the phase plate.

Since the original machine was built we have found much better motors for driving higher inertias and so either larger aperture phase plate or higher linearities or both can now be produced.

6. PROTOYPE INTERFEROMETER

The first attempt to build a visible phase interferometer based on the phase plate method of modulation was deliberately constructed on a wooden desk. This was done for two reasons:-

(1) To demonstrate that a compact desk top design was possible.

(2) The aim was to produce an instrument that did not need expensive vibration isolation and could be used in harsher environments than is common for interferometery. If an instrument could be made that would work on such a base it would prove that capability.

A schematic of the optical layout is shown in fig 5. The interferometer is a Twyman Green with two phase plates one in the test arm and one in the reference arm. From the schematic it can be seen that a HeNe laser with a beam expander is located under the slate slab (used as the optical bench). The expanded beam is reflected up into the interferometric head. Beamsplitter B1 separates the reference and test arms.

The beam in the reference arm traverses a phase plate, which in this case was a glass microscope slide, and is reflected back along its own path by mirror M1 to the beamsplitter. The sample beam is transmitted through a similar phase plate into the test arm which can contain a lens or piece of material for test. The light transmitted by the test sample is reflected back by a suitable mirror M2 to the beamsplitter. The two beams are combined at the beamsplitter and travel through lens L1 onto beamsplitter B2 where part of the light from the interferogram is reflected onto a reference detector. The reference detector is a small area photodiode that looks at one part of the interferogram continuously. The remainder of the beam is focused on to two orthogonal scanning mirrors which scan the interferogram in a TV type raster of 100 lines over a second photodiode. The second detector "looks" at the whole of the interferogram during the course of a complete scan.

The phase plates are mounted on two General Scanning galvanometers (GP 300) and are driven by servo amplifiers. The galvanometer motors both for scanning and phase plate drive have inbuilt position transducers and can therefore by means of a servo system be made to follow a given input signal. The phase plates are driven with the same sawtooth input signal as the linescan galvanometer which keeps the phase plates and the linescan synchronised.

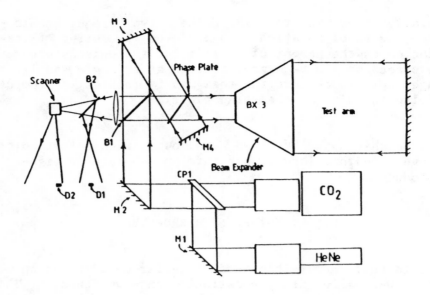

Fig. 8. Schematic of Multiwaveband Interferometer

The optics for the instrument are shown in fig 8 and it can be seen that the reference beam has been reflected to intersect the test arm and where they intersect a single phase plate has been used. It can be seen that as the phase plate is rotated it increases its angle in one beam whilst decreasing its angle to the other beam. This simplifies the instrument by reducing the number of scanning motors and their associated electronics. The single plate makes setting up much easier as there is now no need to keep the amplitudes and offsets of two galvanometers matched since they are now one in the same.

The optics consist of two lasers which are expanded by separate beam expanders up to 25 mm. In the case of the CO_2 LASER the expansion is 20x and in the case of the visible it is 30x. The expanded beams in this dual waveband instrument are incident on a beam combiner plate CP1. The HeNe is reflected onto CP1 by M1. The CO LASER is transmitted through CP1 to combine with the reflected HeNe and then the two beams travel along a common path to M2. From M2 the two laser beams are reflected up through the table to the beamsplitter B1. From the beamsplitter one beam is transmitted into the reference arm which consists of a folding mirror M3 from which the reference beam is reflected through the phase modulation plate P to the reference mirror M4. The reference mirror reflects the beam back along its own path back to the beamsplitter.

In the case of the beam reflected from the beamsplitter this forms the sample beam. The sample beam is also transmitted through the phase plate and normally travels through a beam expander BX3. The standard beam expander is 150 mm. but any good quality beam expander can be used. After the beam expander the beam would normally be transmitted through a test object either a lens or optical material or it would be reflected off a surface to be tested. In the case of a transmitting sample the beam would be reflected back along its own path by a plane or curved mirror.

The phase plates are driven with sufficient amplitude to produce at least 100 wavelengths of phase shift per linescan. With this experimental set-up different methods of driving the phase plates were tried ie. triangular waveforms were used, different mark space ratios on the sawtooth signal were investigated. The triangular waveform confirmed a known error in the system. Results scanning in one direction gave different answers to those in the opposite direction. This was caused by the fact that measurements are taken during scanning. If the interferometer is set-up to have increasing phase in the direction of the linescan then it can be seen that with the phase plate also moving to increase phase the two components are additive and give increased frequency of test signal. If the direction of the phase plate is now reversed the phase plate will be reducing phase at a time when the scan is causing an increase: the two are now subtractive resulting in a lower frequency. This error became known as the Doppler shift error for it is a very similar phenomena. Without correction the triangular scan gave a very confusing picture.

The information from the interferometer needed to be displayed in a form that would make its interpretation easy. The prototype instrument used a BBC computer (uses a 6502 microprocessor) to produce a picture made up of 100 by 100 coloured pixels on the screen. The colour of each pixel was determined by the phase value of the corresponding point in the interferogram. The first pictures produced were noisy, it was found that the phase plates had vibration induced in them by the galvanometers and that the signals from the detectors required better electronic filtering. The mounting of the phase plates on the galvanometers was improved and so were the filters on the electronics.

The prototype demonstrated that the original aims of the instrument could be met. The system produced good interferograms on the wooden desk with no antivibration mounts. Trying to view fringes in the conventional manner by eye showed the fringes to be very unstable and drifting too quickly for assessment. However turning the phase measurement on displayed clear stable fringes. The display speed was limited by the computer speed and the multiwaveband aspect could not be tested due to a lack of suitable optics.

7. FIRST MULTIWAVEBAND INSTRUMENT

The principle differences between this and the first instrument were:-

        (a) Single phase plate used in both arms
        (b) Dedicated processor (Cube system)
        (c) ZnSe optics with multiwaveband coatings

The instrument was built on a honeycomb table top. The interferometric head was machined out of a solid block of aluminium and formed a very compact assembly.

The return beam from the sample arm is adjusted to travel back along its own path to the beamsplitter B1 where it recombines with the reference beam. The two beams travel to the focusing lens L1 which focuses them onto the scanning mirrors via the beamsplitter B2. About 20% of the beams are reflected from B2 onto the reference detector D1 the remaining 80% being scanned by the scanning mirrors in a TV type raster across the sample detector D2.

The detector D1 "sees" only one part of the interferogram and the part that it does "see" can be adjusted by rotating the beamsplitter B2. The adjustment of B2 is designed to deal with, catadioptic lenses, The detector D1 provides a reference signal and an indicator LED on the control panel shows when the reference signal is present and thus indicates when the beamsplitter is in a correct position. The beamsplitter B2 also has the facility to rotate through 90 degrees this enables the interferogram to be viewed by a suitable camera placed on the port on the top off the interferometer. To enable the fringes to be viewed by a camera, either a vidicon or pyrovidicon, the phase plate needs to be switched off as the modulation of the optical paths blur's the fringes.

With the exception of the two LASER beam expanders. The reflecting surfaces are coated with enhanced reflectivity coatings and all the antireflection coatings are designed to operate over the bandwidth of 0.6um to 11um.

8. RESULTS

The new instrument performed very well in both the visible and IR with the following specification:-

Repeatability	$\lambda/32$
Resolution	$\lambda/240$
Number of data points	10,000
Scan Rate	4.5 seconds/frame

Effort was put in to reduce sources of noise, one area that needed improvement was pickup in the electronics and a second more fundamental source was ringing in the aluminium housing of the interferometer. The ringing was caused by vibrations induced by the phase plate drive. Different methods of mounting the motor onto the housing were tried and the noise reduced but not eliminated.

It was decided to design a full commercial version of the interferometer using the present instrument to carry out experiments to try out ideas for the commercial instrument.

The present instrument still produced the Doppler shift error which could be computed out after the data had been captured and displayed. A correction routine was written into the capture programme to try and correct real time but it was too slow for real time operation. It was decided that the commercial instrument should have a faster processor (68020 + floating point chip) in it to overcome the problem. The data captured by the interferometer had discontinuities in at 2PI phase shifts this was necessary to determine if phase was increasing or decreasing. These discontinuities also needed to be taken out with the software and the correct fringe order ascribed to each discontinuity. The commercial instrument deals with the Doppler error and the discontinuities real time. The optics are mounted on synthetic granite which has removed the problem of ringing. The results for the new instrument are:-

Repeaability $\lambda/64$
Resolution $\lambda/320$
No. of data points 10,000
Scan rate   4.5 secs/frame

9. CONCLUSIONS

The phase interferometer stared as an in house instrument only, with a set of objectives designed to meet our own requirements.

The prototype instrument quickly demonstrated that the proposed technique could with development meet the original objectives listed below:-

(a) Deal with complex fringe patterns
(b) Able to be used in a production environment
(c) Faster and easier to use than WATIRL
(d) Multiwaveband ie. (visible and IR)

The prototype multiwaveband instrument verified that all the objectives could indeed be met at this time the project was extended to develop the instrument further into a commercial instrument.

ACKNOWLEDGEMENTS

The author wishes to thank D.M. Ring and G. Cross for their excellent support. This paper is published with the permission of the Directors of Pilkington PLC and Dr. A. Ledwith, Director of Group Research.

References

1. D.R.J. Campbell, Accurate assessment of the optical quality of infrared systems, Optical Engineering March/April, 1984 Vol. 23 No. 2.

2. R.L. Drew., A simplified shearing interferometer, Proc Phys. Soc. B. 1951 Vol. 64

Fringe Modulation for the Separation of Displacement Derivative Components in Speckle-Shearing Interferometry

J. Fang[*] and H. M. Shi[**]

* Department of Mechanics, Peking University, Beijing, 100871, China
** Department of Engng. Mechanics, Tsinghua University, Beijing, 100084, China

ABSTRACT

Both optical information processing and digital image Processing are used for the component separation of the displacement derivatives in speckle shearing interferometry. In recording, a single-beam illuminates normally the object surface and two photographic plate record symmetrically the speckle field. Fringe carriers are introduced during the double exposures, which result in the frequency difference between the in-plane and the out-of-plane information. Through the order-interpolation of the modulated fringe carriers, the in-plane strain fringes are separated by digital-image subtraction.

1. INTRODUCTION

With advantages as simple setup, less vibration isolation, etc. the image-speckle-shearing camera developed by Hung [1] is a convenient system of speckle interferometry for the measurement of displacement derivative fields. The full-aperture utilizing of the camera lens not only reduces the exposure time in recording but also fining the speckles on image plane. However, the full-aperture of large size brings also two problems. One is the displacement modulation to the strain fringes by the displacement components involved in recording result, which has been investigated by the author [2] in another paper in that improving method was proposed. Another problem is the component-separation of the displacement derivatives that is expected to be solved in this paper. Similar to the case in holographic interferometry, the fringe formation of speckle shearing by single aperture relies on both illumination direction and viewing direction. Illuminating normally the object surface one can obtain the pure out-of-plane displacement derivative fringe patterns by recording in the normal to the surface. For the acquisition of the in-plane displacement derivative patterns, however, it usually require two fringe patterns resulting from different illuminating angles to separate the in-plane components point by point. To obtain directly the in-plane strain fringes, Dai [3] et al suggested to use symmetrical double-illuminating so as to eliminate the out-of-plane derivative components by recording of four exposures, in which fringe carriers were introduced to enlarge the frequency-difference between the in-plane and the out-of-plane information, and the in-plane fringes was expected to be read-out by double optical-filtering. Because the derivative fringes consist of speckles it is difficult to obtain modulated carrier fringes whose frequency is large enough to meet the diffraction separation in filtering.

In this paper, the recording of four expoures is replaced by double exposures under single-beam illumination. The optical filtering processing is then combined with the digital-image processing to fulfil the separation of the in-plane displacement derivative fringes.

2. RECORDING AND OPTICAL IMAGE PROCESSING

A beam collimated by lens L is used here to illuminate the object towards the normal to the diffused surface, as shown in Fig. 1. Two lens L_1 and L_2, each of them is covered by a shearing element S, consist of two single-aperture speckle-shearing cameras, that are symmetrical about the normal with angle Q in plane X-Z. The first exposure is made before the deformation of the object, and two films H_1 and H_2 are used to record respectively the intensities of

For H_1
$$I_1'(x,\ y)=a|\exp[A_1(x,\ y)]+\exp[A_1(x,\ y)+\Delta A_1(x,\ y)]|^2 \tag{1}$$

For H_2
$$I_2'(x,\ y)=a|\exp[A_2(x,\ y)]+\exp[A_2(x,\ y)+\Delta A_2(x,\ y)]|^2 \tag{2}$$

where $A_1(x,\ y)$, $A_2(x,\ y)$ are random phases of speckle field, and $\Delta A_1(x,\ y)$, $\Delta A_2(x,y)$ are their increments due to the image-shifting, or

$$A_i(x,\ y) = A_i(x + \Delta x,\ y) - A_i(x,\ y) \qquad (i=1,\ 2) \tag{3}$$

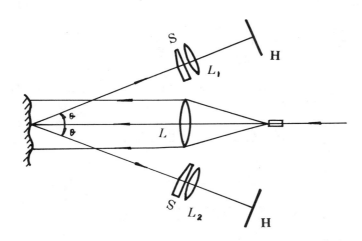

Fig. 1 Schematic diagram of illumination and recordings

After the object is loaded, the phases are changed due to the optical path variation. Before the second exposure is made, the lens L is moved along the optical axis towards the object surface to introduce the fringe carrier. Therefore, the second exposure records the intensity of

For H_1
$$I_1''(x,\ y)=a|\exp[B_1(x,\ y)]+\exp[B_1(x,\ y)+\Delta B1(x,\ y)+2\pi f_o x]|^2 \tag{4}$$

For H_2
$$I_2''(x,\ y)=a|\exp[B_2(x,\ y)]+\exp[B_2(x,\ y)+\Delta B_2(x,\ y)+2\pi f_o x]|^2 \tag{5}$$

The total intensity recorded by H_1 and H_2 are, respectively

For H_1
$$I_1(x,y)=I_1'(x,y)+I_1''(x,y)=4a^2[1+\cos\frac{\Delta B1+\Delta A1+2\pi f_o x}{2}\cos\frac{\Delta B1-\Delta A1+2\pi f_o x}{2} \tag{6}$$

For H_2

$$I_2(x,y)=I_2'(x,y)+I_2''(x,y)=4a^2[1+ \text{Cos} \frac{\Delta B_2+ \Delta A_1+2\pi f_0 x}{2} \text{Cos} \frac{\Delta B_2- \Delta A_2+2\pi f_0 x}{2}] \tag{7}$$

The developed photo graphic plate are optically processed in Fourier transform system. Though each cosine term in Eq. (6) are with basic frequency f_0, the last term cos $[\Delta B_1- \Delta A_1+2\pi f_0 x]/2$ can still be separated from the first one by filtering since it consists of regular factors that related to the displacement derivatives by

$$\Delta B_1- \Delta A_1= \frac{2\pi}{\lambda} [\frac{\partial w}{\partial x}(1+\text{Cos } \theta) + \frac{\partial u}{\partial x} \text{Sin } \theta] \tag{8}$$

By a filtering that lets the rays of the frequency caused by randoms $\Delta A_1+ \Delta B_1$ pass through the back-focal plane of the transform lens, the intensity on the ontput plane $T(x_i, y_i)$ will be:

$$I_{i1}(x_i, y_i) = K_1 \text{Cos}^2 \frac{\Delta B_1- \Delta A_1+ 2\pi f_0 x_i}{2} \tag{9}$$

where K_1 is a constant.

Similarily, because the factor in the last cosine term of Eq. (7) is

$$\Delta B_2- \Delta A_2= \frac{2\pi}{\lambda} [\frac{\partial w}{\partial x} (1+\text{Cos } \theta) - \frac{\partial u}{\partial x} \text{Sin } \theta] \tag{10}$$

the processing of plate H_2 can be made in the same way as that for H_1, and intensity on the image plane will be given by

$$I_{i2}(x_i, y_i) = K_2 \text{Cos}^2 \frac{\Delta B_2- \Delta A_2+ 2\pi f_0 x_i}{2} \tag{11}$$

where K_2 is a constant of intensity.

3. DIGITAL IMAGE PROCESSING

The patterns from the optical filtering are two fringe carriers of basic frequency f_0 modulated by displacement derivatives, as expressed above in Eq. (9) and Eq. (11). Because of the influence of speckle noise and the range limit of the displacement measurement, it is impossible for the frequency f_0 to be high enough to produce moire fringes by the superimposition of the two carriers. To solve this problem, the patterns are imported into digital image processing system for processing. First of all, the fringes of grey images are transformed into binary ones, and then are coded with monotonical fringe orders. Then, a linear interpolation of fringe order is made to increase the 'density' of the fringes. Two digital images corresponding to two fringe carriers are obtained with fringe-orders of

Image 1: $$N_1(x_p,y_p)= \frac{1}{\lambda} [\frac{\partial w}{\partial x} (1+\text{Cos } \theta)+ \frac{\partial u}{\partial x} \text{Sin } \theta+2\pi f_0 x] \tag{12}$$

Image 2: $$N_2(x_p,y_p)= \frac{1}{\lambda} [-\frac{\partial w}{\partial x} (1+\text{Cos } \theta)- \frac{\partial u}{\partial x} \text{Sin } \theta+2\pi f_0 x] \tag{13}$$

where x_p, y_p are coordinates of Pixel $p(x_p, y_p)$.

with the help of image subtraction, we have

$$N(x_p,y_p)=N_1(x_p,y_p)-N_2(x_p,y_p)= \frac{2 \sin \theta}{\lambda} \left[\frac{\partial u(x_p,y_p)}{\partial x} \right] \tag{14}$$

where $N(x_p,y_p)$ is fringe order of the in-plane strain $\partial u/\partial x$. After subtraction, an average smooth could be performed to improve the quality of the subtractive moire fringes.

4. CONCLUSION

The illuminating of single-beam and the recording of double-exposure not only simplify the recording procedure but also avoid the interaction between the information so that the fringe carriers are introduced independently. The combination of the optical information-processing with the digital image-processing is a way to explore advantages of each technique. The digital interpolation of fringe orders increases the density of the carriers and ensures the generation of the in-plane strain fringe patterns.

5. REFERENCES

[1] Hung, Y. Y. and Liang, C.Y. Image-Shearing camera for direct measurement of surface strains, Appl. Opt. 18, 1046 (1979)

[2] Fang. J., Fringe variation and visibility in speckle-shearing interferometry. SPIE 1164, (1989)

[3] Dai, F. L. Wang, S. Y. and Zhong. G. C. 'speckle-shearing moire for strain fields' Proc. of 4th Conf. on Exp. Mech., Wuhan, China. (1984)

Automated laser-diode interferometry with phase-shift stabilization

Yukihiro Ishii

University of Industrial Technology, Department of Electronics
1960 Aihara, Sagamihara, Kanagawa 229, JAPAN

ABSTRACT

A Twyman-Green phase-shifting interferometer with a laser diode (LD) source was constructed for testing an optical element. An automated interferometric system was developed in which the laser current is continuously changed to synchronize intensity data acquisition with vertical drive pulses of a charge-coupled device (CCD). The intensity of interference pattern is integrated with a CCD detector for intervals of one-quarter period of one fringe. A microcomputer calculates the phase to be measured. To avoid the mode instability of LD, a feedback interferometer with extra TTL electronics is made to stabilize the phase shift using the frequency tuning of LD. The experimental result is shown to measure the profile of a diamond-turned Al surface.

1. INTRODUCTION

Recently, laser diodes (LD's) have been proved to be useful light sources in optical interferometry for their excellent features such as a single-mode operation and a frequency tunability.[1] Several authors[2-5] have developed an active heterodyne or a fringe-scanning interferometry with a LD. In phase-shifting interferometer, a piezo-electric transducer (PZT) is commonly used as a frequency shifter. In contrast to using a PZT, a phase measuring interferometry can be possible to be realized by changing the relative phase difference between the two beams of the interferometer with the frequency modulation of LD.

This paper provides a Twyman-Green phase-shifting interferometric system based on the integrated-bucket method[6] with the continuous wavelength change by varying the injection current of LD. With a charge-coupled device (CCD), the phase measurement is automatically performed by synchronizing the fringe data acquisition with CCD clocks. The problem associated with doing the measurement is that the deviations of phase shift from its nominal value due to mode instability of LD and other causes violate the assumption of known phase shift in the phase-extraction algorithm. To avoid this difficulty, a feedback interferometer with TTL electronics is made to stabilize the phase shift using the adaptive frequency tuning of LD. The experimental result for measuring the diamond-turned Al surface is presented.

2. OPERATION OF LD INTERFEROMETER

2.1. Optical system

Figure 1 shows an electronics block diagram of a Twyman-Green interferometer with an unbalanced optical path length 1 such that the phase in this path length can be shifted by changing the wavelength of LD. The light source used in the experiment is a commercially available single-mode AlGaAs LD (HL 7801E) operated at $\lambda_0 = 780$ nm with a 58-mA current and a 20°C temperature of heat sink. The frequency or wavelength is controlled by the injection current such that the temperature must be kept constant. To stabilize the temperature of LD, an automatic temperature-controlled circuit (ATC) together with a Peltier-effect element was used in the experiment. The emitting wavelength of LD with an ATC system is enough stable to ~0.002 nm for the running time from 10 min to 1 hr. The temperature is regulated to ±0.01°C by this system. Therefore, it can be minimized the temperature variation during the experiment.

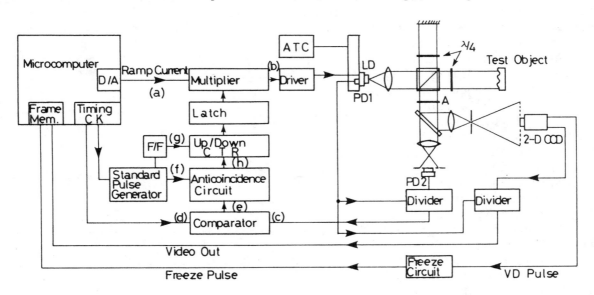

Fig. 1. Experimental setup of a Twyman-Green phase-shifting and feedback interferometer to stabilize a phase shift with a LD.

The emitting wavelength linearly increases over an operating range with no appearance of a mode hop. The laser light collimated by an objective enters a polarizing beam splitter, and light returning from the test surface and the reference surface interferes. The image of test surface is formed on a CCD camera with a magnification of 2 to increase the lateral resolution on the CCD plane. The output power is detected by a monitor photodiode (PD1) to normalize the interference signal on each CCD element and the interferogram detected by a PIN photodiode (PD2) which is fed to a feedback circuit. The interference pattern is read out by a CCD camera (768x493 pixels) whose video signal is converted into an 8-bit digital signal by a frame memory. A microcomuter can calculate the phase extraction routine with 256x256 sample points, and the measured phase is displayed on a x-y plotter.

When the current of LD varies linearly with time from i to i + Δi as Δi= βt, the wavelength changes by $\Delta\lambda$ with the relation $\Delta\lambda=\alpha\Delta i=\alpha\beta t$ where α is a current tuning rate, i.e., 0.0046 nm/mA and β is a constant. The phase shift $\Delta\Phi$ introduced by the wavelength change is

$$\Delta\Phi = \frac{2\pi L\alpha\beta t}{\lambda_0{}^2} = 2\pi ft$$

where the frequency of the heterodyne signal is $f=l\alpha\beta/\lambda_0^2$. In the following experiment, the path difference l and the constant β are given as 30 mm and 300 mA/sec, respectively and then the frequency f is found to be 70 Hz.

According to the integrated-bucket technique, the interference signal of each data set is integrated over the time in which the current varies linearly through a $\pi/2$ shift in phase. The phase to be detected at each data point can be extracted from four integrated signals.

2.2. Feedback interferometer

The accurate measurements[7] require the phase shift $\Delta\Phi$ given by the wavelength change to be stabilized during each bucket of fringe data acquisition. The timing diagram is depicted in Fig. 2 showing how the phase-shift stabilization is accomplished. The alphabets in the figure correspond to those in the specified configuration in Fig. 1. The ramp waveform (a) produced by a D/A converter through a computer is driven to a LD driver circuit. The interference fringe signal (c) divided by a photodiode (PD1) is fed to a comparator which generates a rectangular waveform (e) showing zero-crossing points of a fringe as compared with a sample/hold output (d). Figure 3 is an oscillograph demonstrating the normalization performance of interference fringe intensity. The phase difference (h) between a standard pulse (f) and a falling edge of a rectangular wave (e) indicating a 2π phase shift in a fringe yields through an anticoincidence (exclusive OR) circuit. The pulse train counter circuit operated at the width of phase difference (h) is designed to trigger an up/down counter when a 2π zero-crossing point proceeds or retards with respect to a standard pulse. This either increments or decrements the modulation voltage V_m (b) through a latch which is applied to a LD driver so that the modul-

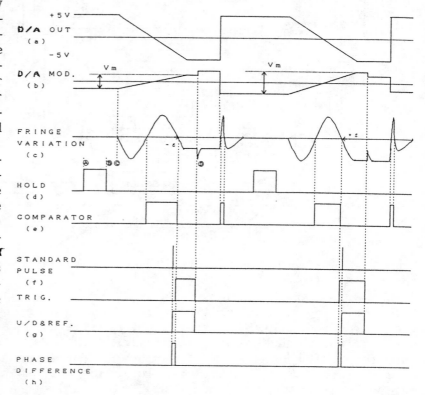

Fig. 2. Timing diagram for the phase-shift stabilization work. The alphabets correspond to those in the experimental setup of Fig. 1.

ation voltage is changed to reduce the width of phase difference (h). The feedback system then is made to lock the phase shift to be zero-crossing point of 2π phase. The experiment demonstrates almost the stabilization characteristics up to ~100 Hz.

Figure 4 shows the phase-shift stabilization work when the feedback loop was on (b) , and off (a) and (c), respectively.

Afterward, by taking four consecutive frames of the fringe intensity data changed by the dc bias of LD current through a latch memory instruction, the integrated-bucket technique is performed. The frame memory in synchronism with CCD clocks[8] reads the intensity data whose integration time is 1/30 sec at a rate of 1/60 sec/frame. The phase to be detected is calculated.

3. EXPERIMENTAL RESULT

The experiment is carried out to verify the capability of the interferometry with a LD. The experiment is performed to measure the surface profile[9] of a diamond-turned

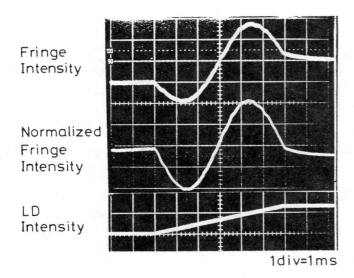

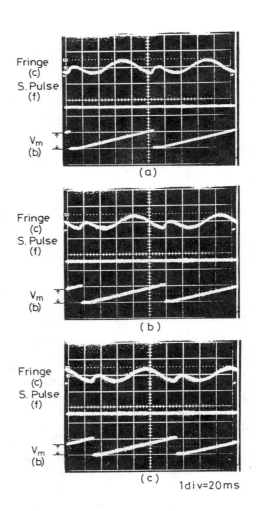

Fig. 3. Oscillograph showing the fringe intensity (above), the normalization performance (middle) and the LD light intensity (below).

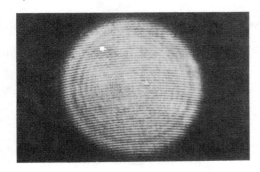

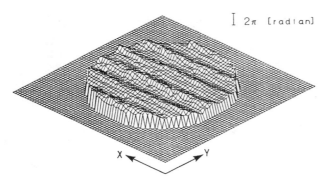

2π [radian]

Fig. 5. The figure above is an interferogram of the surface of a diamond-turned Al disk. The figure below is a 3-D plot of the surface roughness.

Fig. 4. Oscillograph shows that the feedback loop is on (b) and off, in which the standard pulse (S. Pulse) proceeds (a) or retards (c) on a 2π zero-crossing point of fringe, respectively.

aluminum disk. The interferogram of the test surface is shown in Fig. 5 and the measured 3-D map is presented in the below figure. The measured height on this surface roughness is ~ 0.26 μm.

The rms accuracy of the present phase-measuring system with a phase-shift stabilization is ~ 0.12 rad which is reduced to $\lambda/105$ whose estimation is performed by measuring repeatedly 5 times on the same sample.

4. CONCLUSIONS

An automated laser-diode interferometric system was developed through the direct modulation of LD wavelength by synchronizing the fringe data acquisition with the CCD clocks. A CCD camera was also employed to integrate the time-varying interference intensity distribution. A feedback interferometer with analog and digital electronics is constructed to stabilize the phase shift by using the frequency tuning of ramp current of LD. A measurement repeatability of $\lambda/105$ was obtained.

Unlike conventional techniques, there is no large and heavy mirrors such as a PZT in the interferometer.

5. ACKNOWLEDGMENTS

The author wishes to thank Mr. Y. Omoto for his experimental help.

6. REFERENCES

1. T. Yoshino, M. Nara, S. Mnatzakanian, B. Lee and T. Strand, "Laser diode feedback interferometer for stabilization and displacement measurements," Appl. Opt. 26(5), 892-897(1987).
2. Y. Ishii, J. Chen and K. Murata, "Digital phase-measuring interferometry with a tunable laser diode," Opt. Lett. 12(4), 233-235 (1987).
3. K. Tatsuno and Y. Tsunoda, "Diode laser direct modulation heterodyne interferometer," Appl. Opt. 26(1), 37-40 (1987).
4. J. Chen, Y. Ishii and K. Murata, "Heterodyne interferometry with a frequency-modulated laser diode," Appl. Opt. 27(1), 124-128 (1988).
5. T. Kubota, M. Nara and T. Yoshino, "Interferometer for measuring displacement and distance," Opt. Lett. 12(5), 310-312 (1987).
6. J. C. Wyant, "Use of an ac heterodyne lateral shear interferometer with real-time wavefront correction systems," Appl. Opt. 14(11), 2622-2626 (1975).
7. P. Hariharan, "Phase-stepping interferometry with laser diodes: effect of changes in laser power with output wavelength," Appl. Opt. 28(1), 27-29 (1989).
8. K. N. Prettyjohns, "Charge-coupled device image acquisition for digital phase measurement interferometry," Opt. Eng. 23(4), 371-378 (1984).
9. O. Sasaki and H. Okazaki, "Sinusoidal phase modulating interferometry for surface profile measurement," Appl. Opt. 25(18), 3137-3140 (1986).

Fringe analyzer for a Fizeau interferometer

Kenji Yasuda, Ken Satoh, Masane Suzuki

Fuji Photo Optical Co., Ltd. Optical Research Dept.
1-324 Uetake-cho Omiya, Saitama, 330 Japan

and

Ichirou Yamaguchi

The Institute of Physical and Chemical Research,
Wako, Saitama, 351-01 Japan

ABSTRACT

We have developed a new computerized measuring system for interferometric testing of lenses and mirrors. A personal computer with an image processor controls a laser interferometer, stores interferograms, derives the wavefront shape, and evaluates such imaging performances of optical components as the Seidel and the Zernike aberration coefficients, the peak to valley(P-V), the root mean square(RMS) values of wavefront aberration, the spot diagrams, PSF, and MTF.

1. INTRODUCTION

The history of optical interferometers and its applications are very long. However, they had not been so widely applied for lens testing in the optical industry before several years ago. Reasons for this are the complexity in the adjustment of interferometers and in evaluation of interferograms. Recently, there have been several automatic interferometric testing systems which are commercially available. Nowadays interferometric testing is required in many fields in industry, for example, in high precision mechanics and optoelectronics. We have to inspect a great numbers of high precision products such as video heads and optical pickups. The conventional manual inspection cannot be applied in this case. Thus automatic high precision measurement of optics in short time is strongly required. We have developed an automatic and quick system for this purpose.

2. OUR SYSTEM

Our system consists of a laser interferometer, and an image processor and a personal computer controlling them. Figure 1 shows the whole system of hardware.

2.1 Interferometer

Our interferometer is a Fizeau interferometer as shown in Fig.2. A flat or a sphere for reference surfaces can be attached to the main body of the interferometer which is framed in the figure.

The interferometer has the special feature that the holder of the reference flat (or sphere) can be rotated. We can measure the components supported both horizontally and vertically. Interference fringes are generated by the combination of the wavefront reflected from the high precision reference surface with that reflected from a surface under test.

We adopted two analyzing methods to get the phase distribution from the interferograms. One is the fringe scanning method in which the reference surface is moved to shift the fringes and the phase values are derived at every point of the interferograms. The other is the fringe center acquiring method in which centers of fringes are extracted by image processing. In the both cases the interferometer is controlled in the following way in order to derive the desired wavefront precisely.

2.1.1 Automatic brightness adjustment

The brightness of the interferograms depends on the magnification of the fringe projecting system, reflectivity of sample surfaces, and output power variation of a laser. In order to avoid the saturation of the video camera, an automatic brightness adjuster composed of a rotatable polarizer is built in the system. It is especially important for the fringe scanning method where sinusoidal intensity variation of fringes is absolutely necessary. The adjuster can also be controlled manually by pushing keys of the computer when the sample has high reflectivity only in a very small area. In this case, the automatic adjuster limits the dynamic range of the other area of the interferograms and consequently damages measurement accuracy. In this case, only isolated pixels at which brightness is saturated or unchanged during fringe scanning are removed from the interferograms by a software.

2.1.2 Turbulence check

The interferometric measuring system should be used in the stable condition. Instability in the fringe is caused by the following factors.
1) Nonuniformity of ambient temperature.
2) Vibration of the floor.
3) Change of laser wavelength caused by the thermal deformation of the laser resonator.

In our fringe scanning method, the fringe phase is basically shifted from the initial state through $\pi/2$, π, $3\pi/2$. In addition to four necessary interferograms, we also store $4\pi/2$ shifted interferogram, which is compared with the initial interferogram for checking stability of the interferometer.

2.1.3 Calibration of piezo element

We adopted the piezo element to shift the fringes. However, our interferometer has the special feature that its reference surface holder can be rotated for measuring optics supported both horizontally and vertically. Therefore, mechanical load added to the piezo element changes in the both cases whereas the expansion coefficient of the piezo element depends on the added load and the ambient temperature. These factors must be compensated for because we adopted a piezo element which has a built-in load cell and is free from hysteresis. In spite of it, a small error may occur. In order to suppress these possible errors, we developed a calibration program for piezo element which is similar to stability check.

In this calibration procedure the initial interferogram is first stored and then a voltage is applied, which is expected to bring about 2π phase shift in the piezo element. Then another interferogram is stored. The voltage is changed gradually until the initial and 2π shifted interferograms become coincident with each other. Thus the correct voltage to cause 2π phase shift is found, and 1/4 of that voltage is applied for shifting $\pi/2$ phase.

2.2 Image input and output

Interferograms taken by a CCD camera of the interferometer are stored in the frame memory ($512 \times 480 \times 8$ bit) of the image processor FX-15. The processor delivers an external synchronous signal based on EIA standard to the CCD camera. The stored image is displayed on the CRT of the image processor. The area of measurement and the menus to be selected by a mouse are also superimposed on it.

2.3 Computer

As a controller of the image processor, we adopted a 32 bit MPU which has enough accessible memory area to process a great amount of image data. Its software was written in an assembly language. The processor grabs interferograms and calculates the wavefront from a component under test. To display the results of measurement we chose a personal computer NEC PC-9801 for the terminal. Storage of interferograms, data analysis, display of the results, and data recording are executed sequentially in an on-line process. On the other hand, we can also analyze wavefronts data stored by measurement in the past by an off-line process and inspect the measured results graphically. The software also includes the utilities for formatting a data storage disk and copying files. These are shown in Figs.3 and 4. The on-line process has a modular architecture.

3. METHOD OF ANALYSIS

In Fig.5 the flow chart of the whole analysis is shown. It consists of the following items.

3.1 Aberration functions

3.1.1 Zernike polynomials

First, we computed the Zernike aberration coefficients of the measured wavefront. The Zernike polynomials are orthogonal over the circle of unit radius. However, we cannot use this relationship because our measuring points are discrete and the aperture is not generally circular. So the distribution of the measured phase is fitted to the Zernike polynomials by the least square method. We represent the Zernike polynomials by the series

$$W_z = Z_0 + Z_1\rho\cos\theta + Z_2\rho\sin\theta + Z_3(2\rho^2-1) + Z_4\rho^2\cos 2\theta + Z_5\rho^2\sin 2\theta + Z_6(3\rho^2-2)\rho\cos\theta$$

$$+ Z_7(3\rho^2-2)\rho\sin\theta + Z_8(6\rho^4-6\rho^2+1) \tag{1}$$

Then we denote the phase value of each point of the measured wavefront by W_{ij} and solve the simultaneous equations to derive each coefficient of the Zernike polynomials.

3.1.2 Seidel's aberrations

The wavefront aberration is fitted to the nine terms of the Zernike polynomials in Eq.(1) by the least square method. The result is converted to the expansion of the Seidel's third order aberration by means of the relationship.

$$W_z = \underbrace{W_0 - Z_3 + Z_8}_{} + \underbrace{\sqrt{(Z_1-2Z_6) + (Z_2-2Z_7)^2}\,\rho\cos(\theta-\alpha)}_{\text{tilt}} + \underbrace{(2Z_3-6Z_8)\rho^2}_{\text{defocus}} + \underbrace{6Z_8\rho^4}_{\text{sph.abb.}} + \underbrace{3\sqrt{Z_6^2+Z_7^2}\,\rho^3\cos(\theta-\beta)}_{\text{coma}}$$

$$+ \underbrace{\sqrt{Z_4^2+Z_5^2}\,\rho^2\cos(2\theta-2\gamma)}_{\text{astigmatism}}, \tag{3}$$

where

$$\alpha = \tan^{-1}\left\{ \frac{(Z_2-2Z_7)}{(Z_1-2Z_6)} \right\} \qquad \text{:azimuth of tilt} \tag{4}$$

$$\beta = \tan^{-1}\left\{ \frac{Z_7}{Z_6} \right\} \qquad \text{:azimuth of coma} \tag{5}$$

$$2\gamma = \tan^{-1}\left\{ \frac{Z_5}{Z_4} \right\} \qquad \text{:azimuth of astigmatism} \tag{6}$$

Thus the Seidel's aberration function has been derived and the Seidel's coefficients and the azimuth angles are deduced from the fitted Zernike polynomials.

3.1.3 PV and RMS values of wavefront aberration

The maximum intensity of the point image depends on the mean square of wavefront aberration $<\Delta W^2>$ that is defined by

$$<\Delta W^2> = \frac{1}{M}\sum_{i,j}(W_{ij}-<W>)^2 \tag{7}$$

with

$$<W> = \frac{1}{M}\sum_{i,j}W_{ij}, \tag{8}$$

where M is the total number of sample points. The maximum intensity is given by

$$i(p) = 1-\left(\frac{2\pi}{\lambda}\right)^2<\Delta W^2>. \tag{9}$$

The position of the maximum intensity does not coincide with the paraxial focus when the aberration is present. In this case, certain amounts of tilt and defocus are added to the wavefront so that the mean square deviation of the wavefront aberration can take the minimum. That is, if $W(p,\theta)$ is the phase distribution measured, we derive the values of H, K, L and M that makes the root mean square $<\Delta W'^2>$ of

$$W'(\rho,\theta) = W(\rho,\theta)+H\rho^2+K\rho\sin\theta+L\rho\cos\theta+M \tag{10}$$

to be minimum. The minimum RMS value of this fitted wavefront is also calculated.

3.2 Imaging performances

Using the aberration data which have been evaluated above, we compute the following diagrams and functions indicating the imaging performance of the system.

3.2.1 Spot diagram

We divide the exit pupil into a rectangular mesh and trace a ray going through each of the mesh points on to the focal plane. In the absence of the aberration, all the rays are focused at a point in the focal plane. Otherwise, the rays are scattered more or less at the plane.

If the wavefront is flat, the rays converges at the optical axis. If the wavefront has a tilt component, the ray is tilted because the ray propagates in the direction normal to the wavefront.

Lateral aberrations K_x, K_y are calculated from the wavefront aberration using the relationship

$$K_x = R\frac{\partial W}{\partial \xi} \quad \text{and} \quad K_y = R\frac{\partial W}{\partial \eta}, \tag{11}$$

where the coordinate of the pupil plane is (ξ,η) and the distance from the exit pupil to the image plane is R.

3.2.2 Point spread function and MTF

We denote the exit pupil by the coordinate (ξ,η) with the pupil function $H(\xi,\eta)$. We also take the coordinates x,y in the image plane and the point spread function h(x,y). They are related to each other through the relation

$$h(x,y) = C\left| \int\int H(\xi,\eta) \exp\left[-\frac{2\pi i}{\lambda R}(x\xi+y\eta)\right]d\xi d\nu \right|^2 \tag{12}$$

where C is a constant. The pupil function is represented as

$$H(\xi,\eta)=A(\xi,\eta) \exp\left[\frac{2\pi i}{\lambda}W(\xi,\eta)\right], \tag{13}$$

where $A(\xi,\eta)$ is the amplitude transmission function and $W(\xi,\eta)$ is the aberration function in the wavelength unit.

The optical transfer function OTF is defined as the Fourier transform of PSF, that is,

$$T(s,t) = \frac{\int\int_{-\infty}^{\infty} h(x,y) \exp[2\pi i(sx+ty)]dxdy}{\int\int_{-\infty}^{\infty} h(x,y)dxdy} \tag{14}$$

The modulation transfer function MTF is derived as the absolute value of the OTF. As shown above, the PSF is calculated from the measured aberration function. The extent of the PSF indicates the imaging acuity of a lens. Our software displays the degree of the extent graphically.

3.2.3 Encircled energy

The encircled energy, defined as the total energy included within a circle of the radius a, is calculated from the PSF the polar coordinate $h(\rho,\theta)$, by using the formula.

$$E(a) = \frac{\int_0^a \int_0^{2\pi} h(\rho,\theta)\rho d\rho d\theta}{\int_0^{\infty} \int_0^{2\pi} h(\rho,\theta)\rho d\rho d\theta} . \tag{15}$$

The origin of the polar coordinate is taken at the maximum of $h(x,y)$. The encircled energy also clearly indicates the concentration of the point image.

4. RESULTS OF MEASUREMENT

4.1 Example of analysis

Figure 6 shows an example of the results obtained from our system. The sample is an optical flat. Its interferogram is shown in the upper left. Only tilt component has been removed. In Output 1, the original interferogram and PV and RMS values are superimposed. Output 2 shows the shape of wavefront. We have also displayed PV and RMS values of wavefront, the contour map and bird's eye view of wavefront. Output 3 is the numerical map of wavefront. Outputs 4 and 5 show the local slope of wavefront in X and Y direction. They have been derived as the phase difference between neighboring points.

In Fig.7 the continuation of outputs in Fig.6 is shown. Output 6 represents the spot diagrams. Because of large aberration, the spots reveal rather broad and uniform distribution. Output 7 shows the wavefront aberration. The PV and RMS values of wavefront aberration, the contour map, and the bird's eye view of wavefront aberration are displayed. The Seidel's third order aberration is also displayed. In Japan, this output scheme is often used because lenses for digital audio discs(DAD) are inspected by measuring the third order aberration coefficients. Thus this figure includes all the necessary items for inspection.

Output 8 shows the coefficients of the Zernike polynomials and those of the Seidel's third order aberration.

Output 9 shows the point spread function. The contour map of intensity is displayed. The cross-section of the intensity distributions along 0, 45, 90 degrees directions are displayed in different colors. In the upper right, diameters corresponding to specific intensity are tabulated. The point spread of this sample shows the sample to be of a low quality.

Output 10 shows the encircled energy. The graph of the encircled energy and values of energy which is included within a circle of the specified diameter are displayed. In this sample energy cannot be concentrated at the center.

Output 11 shows the MTF. The MTF curve and MTF values along 0, 45, 90, 135 degrees directions are displayed. This sample has low resolving power. It forms consequently low contrast images. Output 12 is a MTF table.

We analyzed the same sample with the power removal, which is equivalent to change of the reference surface shift of focus. The result of analysis is shown in Fig.8. The spot diagram shows that rays are concentrated around the center in Output 1. We can notice that PV and RMS values become smaller in Output 2. The point image is also concentrated about the center in Output 4. Energy concentration is improved in Output 5. MTF is closer to the diffraction limited case in Output 6.

4.2 Examination of accuracy

We examined the accuracy of our analyzer in the following.

4.2.1 Aberrations

For evaluating the accuracy of PV value of the wavefront aberration, we calculated PV value from manual analysis of the interferogram and then compare the result of the automatic analysis with that of the manual analysis. The error proved to be less than $\lambda/20$. For evaluating the accuracy of the RMS value, we compared the results with those obtained from other instruments. We found good agreement with the error less than $\lambda/100$. The repeatability of our system has been found to be $\lambda/1000$ in r.m.s.. The numerical values of the aberration functions are also compared with the results of a computer simulation. The error was less than 2% and we compared the results with those obtained from other instruments and the error was less than $\lambda/20$ in the peak-to valley and $\lambda/100$(RMS).

4.2.2 Imaging performances

In order to examine the program for evaluating PSF and MTF, we compared the results for the diffraction limited imaging systems, which have rectangular and circular apertures, with the values computed from the analytical relations. The agreement was better than 1%.

4.3 Processing time

It took 60 seconds to carry out the whole analysis which consists of storage of images, measurements of wavefront shapes, fitting to the aberration functions, and calculation of imaging performances.

5. CONCLUSION

We have developed a high speed automatic fringe analyzer which controls a Fizeau interferometer, stores interferograms, analyzes the wavefront, and evaluates the imaging performances such as aberrations, spot diagrams, PSF, and MTF within 60 seconds.

Repeatability of the aberration analysis is $\lambda/100$ for PV value and $\lambda/1000$ for RMS value. The whole analysis is carried out fully automatically without special skill and knowledge of operators. It can be used for mirrors and lens having diameter between 5 mm and 100 mm and the F-number between F/0.7 and F/15. We are now trying to continue to examine the accuracy of the analyzer, measuring other types of optical components and comparing the results with those from other measuring techniques.

6. ACKNOWLEDGMENTS

The authors thank M. Yoneda, M. Nakata, and M. Kanaya for experimental help and supports.

7. REFERENCES

1) M. Born and E. Wolf: Principles of Optics, Pergamon Press, (1980)459.
2) D. Malacara: Optical Shop Testing, John Wiley and Sons, (1978)489.

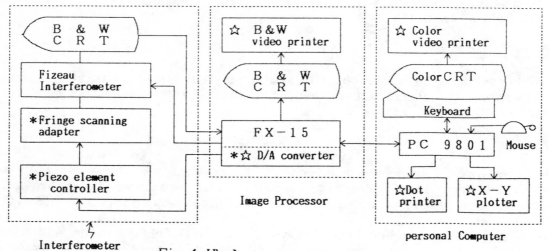

Fig.1 Whole system of hardware

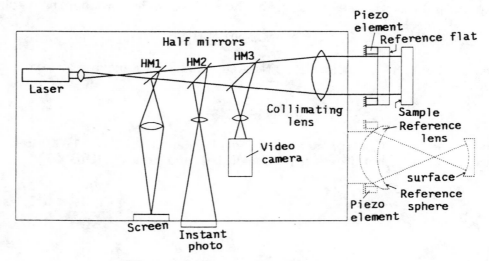

Fig.2 Fizeau interferometer

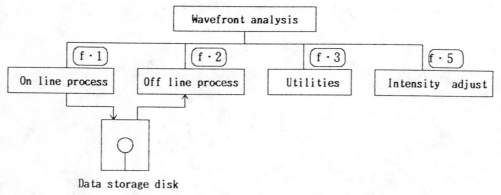

Fig.3 Composition of software

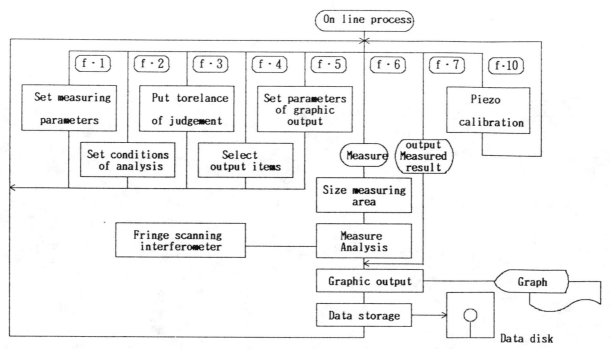

Fig.4 Function diagram of on-line measurement process

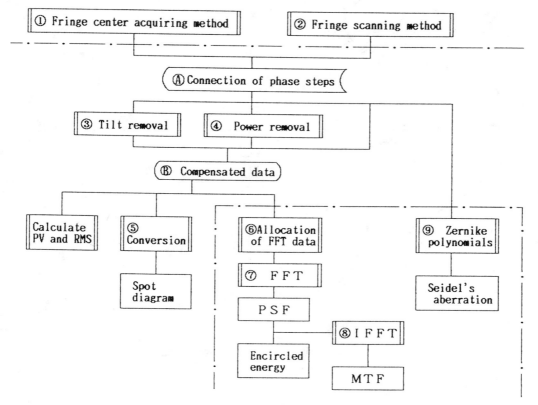

Fig.5 Function diagram of analysis software

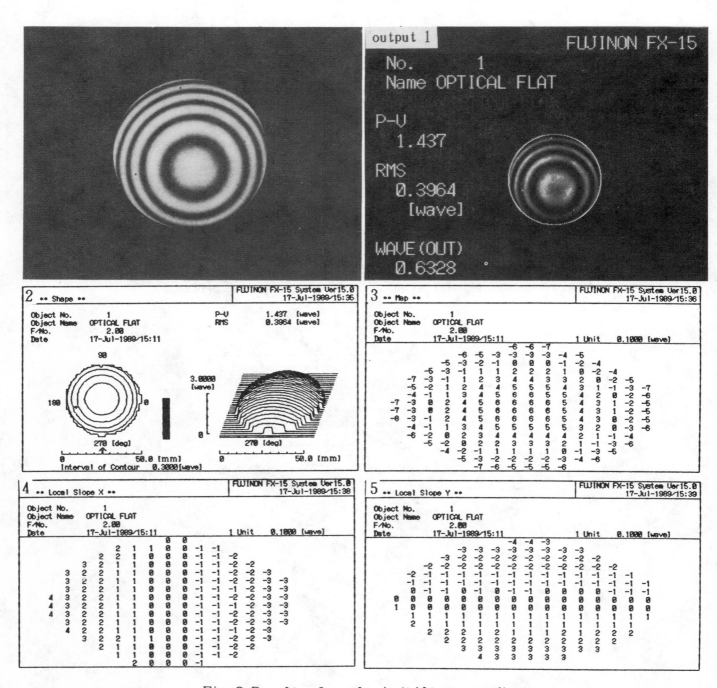

Fig.6 Result of analysis(tilt removed)

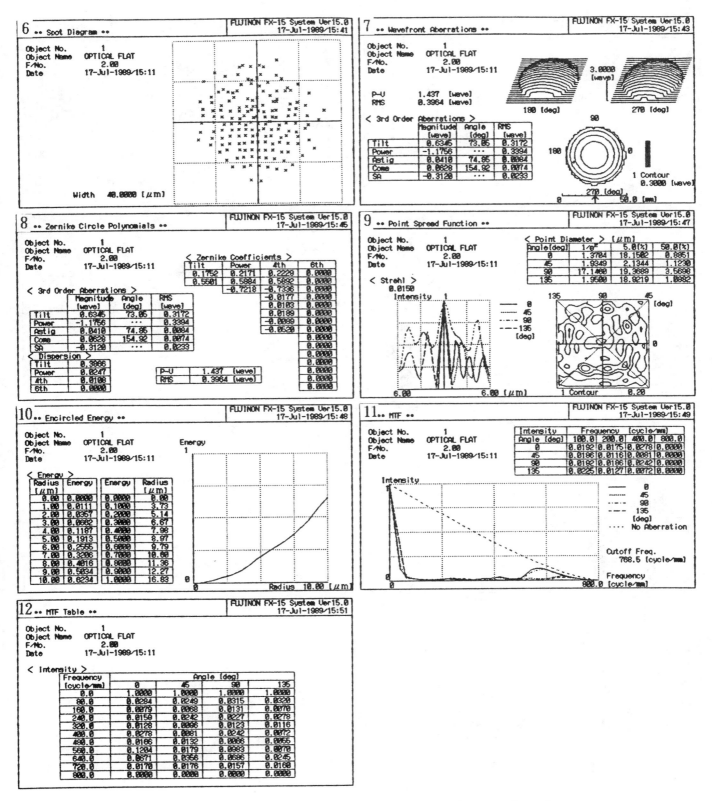

Fig.7 Result of analysis(tilt removed)-continued

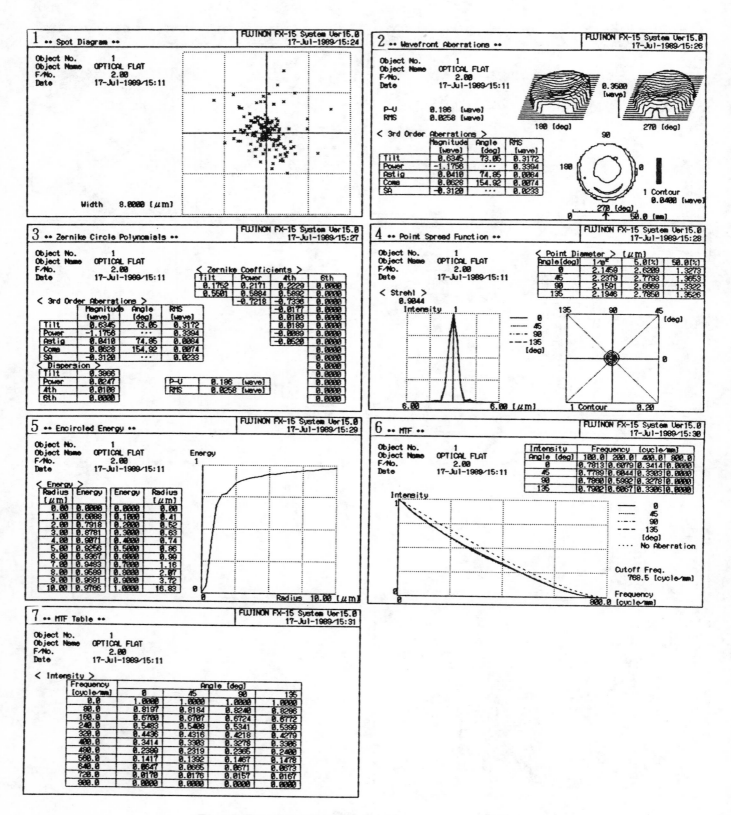

Fig.8 Result of analysis(power removed)

SESSION 4

New Developments and Applications

Chair
Kjell J. Gåsvik
Norwegian Institute of Technology/SINTEF (Norway)

Automatic Video Inspection Of A Diesel Spray

V. K. H. Cheung

T. R. Judge

P. J. Bryanston-Cross

Engineering Department Warwick University

1.0 Introduction

1.1 The Diesel Spray Jet

The subject of the inspection is a diesel spray jet nozzle (fig 1.). The jet is manufactured to spray diesel fuel in 8 precisely controlled directions inside a diesel engine. The angle of elevation of the jet as it leaves the nozzle and the angle between neighbouring jets about the nozzle are of particular importance.

1.2 Specification

The system was initially conceived as a pass/fail test for the production line with a specified inspection time of around two seconds. In this time the system must determine whether the jets are within the specification. This stated that the jets should be sprayed accurately to 0.5 of a degree in elevation and rotation.

1.3 Speed Priority

The two second window for the inspection was a stiff specification, considering that processing was to be carried out on an IBM PC AT. The classical approach to this type of problem would involve processing the entire video field to extract jet features. This kind of processing involves the throughput of a large quantity of data (fig 2.). It is time consuming and certainly could not be carried out within the time frame allowed. To complete the feature extraction within a reasonable time period a more direct approach has been evolved. This relies upon knowledge of the object. It enables the amount of data processed to be substantially reduced and leads to a corresponding decrease in processing time, which allows the specification to be met.

1.4 Inspection Of Sprays

The technique involves the use of cameras to visualize the diesel jets as they leave the nozzle. The jets are seen as diesel is pumped through a sample nozzle under low pressure. The lighting is carefully arranged so that the jets light up brightly like light pipes (figs 1 & 3). The images from the cameras are captured using a 1/25 second frame grabber with four frame stores at a resolution of 512 by 512 pixels.

1.5 Nozzle Description

According to the product specifications, there are 8 equally spaced jets at an elevation of 18 degrees from the horizontal plane. It is assumed that the jet spray is straight. (This is a fair assumption when the part of the jet viewed is close to the nozzle).

1.6 System Geometry

A three camera configuration was found necessary to resolve all angles to sufficient accuracy (figs 4 & 5). The system requires one overhead camera and two auxiliary cameras mounted at 45 degrees to the horizontal, and at right angles to each other.

2.0 Processing Approach

2.1 Computer Model

The basic principle involves comparison of the views of the jets as seen through each camera with a computer generated model of the jets viewed from the same perspective.

2.2 Features Of Model

The inspection requires the extraction of a few important features of the jet configuration so that a comparison may be made. The computer model allows the identification of a set of key points in the video frames that, when examined, will make the pass/fail test of a jet possible.

2.3 Detection Of Jets

A basic requirement is that the jets should be distinguishable from their background. The lighting achieves this, producing an almost binary image. The jets are thus easily detectable, and found above a certain threshold. This threshold is found by averaging one hundred random pixel samples from the frame being considered.

2.4 Prediction Of Jet Position

The rotational orientation of the nozzle to the overhead camera is undefined. Moreover the camera has no point of reference from which it may find this orientation. The problem is resolved and the orientation found by use of a jet template.

The template may be imagined as shown in (fig 6). It has eight arms, one for each jet. This template is notionally placed over the image from the camera. Ten pixels along each arm are sampled, and the number of samples above threshold gives a confidence for whether a jet was at that position.

The template is rotated over the image in steps of 0.5 degrees until it finds the best fit. It records this angle and uses it as a reference. This allows the system to predict where the jets should appear in the views from the auxiliary cameras and detect error.

3.0 Mathematical Computations

3.1 Stages Of Computation

i) Definition of a coordinate system to define the jets' positions in 3D. This is termed the Master Coordinate System. It is a right handed 3D coordinate system, its origin is the centre of the jet nozzle and the z axis extends from it towards the overhead camera.

During the inspection some pixel points in the frames of the auxiliary cameras must clearly be sampled. The positions of these sample points are selected on the model in the 3D master system. To obtain the pixel coordinates of the points within each video frame two transformations of coordinate system are necessary for each auxiliary camera.

ii) The orientation (δ) of the jet set with respect to the overhead camera is found via the template.

iii) The master coordinates of the points are geometrically transformed to the eye coordinates of each camera.

Each camera has its own eye coordinate system. These are left handed 3D coordinate systems, with origins at the centre of the CCD within each camera and z axis extending towards the nozzle. The transformation from the master system is performed to make perspective projection easier. The rotation of the jets (δ) is taken into account in the transformation.

iv) by a perspective projection into coordinates relevant to each video frame.

3.2 Master Coordinate Definition Of Computer Model

Assuming each jet to be of unit length, the master coordinates of the eight endpoints of the jets can be determined by the following formula.

$$x = \cos\theta \, \cos\phi$$

$$y = \cos\theta \, \sin\phi$$

$$z = \sin\phi$$

θ is the angle of elevation of a jet (in this case, $\theta = 18$ degrees) and ϕ is the rotational angle of a jet with respect to the overhead camera. The positions of the endpoints are tabulated in (fig 7).

3.3 Geometric Transformation Between Master And Eye Coordinates

A geometric transformation T converts the master coordinates to eye coordinates. The steps in the process are shown (figs 8 to 11). The eye indicates the camera position, v is the distance of the camera from the nozzle, α is the rotational angle of the camera in the x/y plane with respect to the master coordinate system, and β is the elevation of the camera with respect to this x/y plane.

Figure 9 shows the translation to the new origin of the system through the distance v. Figure 10 shows a rotation of 90+α about the ze axis, and figure 11 shows a rotation of 270-β degrees about the xe axis.

There is one final rotation of the coordinate system that takes into account the rotation of the jet sprays with respect to the overhead camera, the need for this was explained in section 2.4. The rotation is by an angle δ about the z axis of the master coordinate system. In addition a scaling factor is supplied to expand the unit length of the jets, this factor is set to 1 as described in section 4.1. The final geometric transformation matrix is shown below.

$$
T = \begin{bmatrix}
\dfrac{(-\sin\alpha\cos\delta + \cos\alpha\sin\beta\sin\delta)}{s} & \dfrac{(-\sin\alpha\sin\delta + \cos\alpha\sin\beta\cos\delta)}{s} & -\cos\alpha\cos\beta & 0 \\
\dfrac{(\cos\alpha\cos\delta + \sin\alpha\sin\beta\sin\delta)}{s} & \dfrac{(-\cos\alpha\sin\delta + \sin\alpha\sin\beta\cos\delta)}{s} & -\sin\alpha\cos\beta & 0 \\
-\cos\beta\sin\delta & -\cos\beta\cos\delta & -\sin\beta & 0 \\
0 & 0 & v & 1
\end{bmatrix}
$$

This same transformation may be used for each of the cameras by substituting the appropriate angles and distances.

3.4 Perspective Projection

The 2d coordinates of the expected position of the diesel sprays within a video frame are computed by performing a perspective projection of the points from their eye coordinates. The subscript v denotes coordinates in the viewing plane.

$$
x_v = \frac{x_e}{z_e}
$$

$$
y_v = \frac{y_e}{z_e}
$$

4.0 Processing At Speed

To enable processing to be carried out at speed extensive use is made of look up tables to avoid the complex arithmetic of section 3 during the inspection phase.

4.1 Polar Coordinates And Look Up Tables

A 2d point may be specified in polar coordinates by angle and radius. The points to inspect for a given jet position may be wholly defined by just specifying the angle as the radii parameters may be computed. The radii parameters are selected by first calculating the point at which a jet at the specified angle would clip the edge of the frame, and then by sampling at ten equally spaced points up to this radius.

For each camera there are two look up tables of different types (3 * 2 = 6 look up tables).

The first type relates an angle (0 to 360 in steps of 0.5) to the frame coordinates (xv, yv) of ten sample points, calculated via clipping as described above, that might be used to detect a jet at that angle in the frame. These tables make use of the nozzle origin position, which is set during calibration.

The relationship is as shown

$$\alpha_v = \tan^{-1}\left[\frac{y_v}{x_v}\right]$$

It may be seen that ze has been cancelled out in this calculation. The parameter v only appears in ze and so for simplicity is set to 1. The scaling factor s is also cancelled out and this is set to 1 also.

The inspection procedure is formulated as a list of interrogations. e.g Is there a jet at such and such a position? The position at this level is set in terms of jet rotation (ϕ) ranging from 0 to 360, and jet elevation (θ) ranging between 13 and 23 degrees, (positions from -5 to +5 degrees around the expected jet elevation of 18 degrees are also interrogated to find the extent of the deviation).

The second type of look up table, therefore, makes use of the T transformation matrix to relate desired inspection positions, in terms of ϕ and θ, to the actual rotational angle that the jet would appear at within the frame (which is the polar angle that is argument to the first type of look up table).

The form of the second type of table is slightly different between the overhead and auxiliary cameras as the elevation part is of no relevance to the overhead camera, elevation is not discernible from overhead. For the auxiliary cameras the table has entries for ϕ in the range 0 to 360 degrees in steps of 0.5 degrees and θ ranging from 13 to 23 in steps of 1 degree.

5.0 Calibration

The system requires that the overhead camera be set normally to the x/y plane, and that the auxiliary cameras are angled at 45 degrees to this plane. In addition each camera must be informed of the position of the centre of the nozzle. The calibration is performed interactively with the computer. A cross is displayed on the monitor and moved to the nozzle origin, the angular calibration is made by sighting the cameras along special bars that slot into the test rig base. The look up tables are then calculated taking into account the position of the centre of the nozzle as indicated by the cross in each cameras view.

6.0 Increasing Accuracy

6.1 Lighting

The plot (fig 12.). shows a comparison between illumination of a jet under the rig's spot lights and under the diffuse lighting of the laboratory. As can be seen the jet is easily distinguishable under the spot lights. The centre of the peak is identified as the jet's position.

6.2 Jet Profile

It may be seen from (fig 12.) that pixel intensities around the jet rise to a peak and fall away in something like a gaussian manner. It is therefore possible to increase the system's accuracy (once the approximate jet positions are found) by a careful examination of the jets profile in terms of these pixel intensities. The peak of this profile over a set of cross sectional cuts could lead to an order of magnitude increase in the accuracy of detection of the jet position. It would also be possible to detect errors resulting from the bore hole of the spray not intersecting the centre of the nozzle. That is, where the bore hole is not only misplaced in rotation, but also translated.

7.0 Conclusion

The jets' positions may be located in three dimensions to an accuracy of plus or minus 0.5 degrees. The feature processing time using an IBM PC AT compatible has been achieved within the 2 second processing period allowed for the operation.

Having performed a three dimensional identification of the diesel jet, it has been shown how an increase in accuracy can be achieved by performing an interrogation of the grey level profile of the jets.

REFERENCES

1. W.M. Newman, R.F. Sproull, "Principles of Interactive Computer Graphics",McGraw-Hill,1981,pg 333-344

2. D.J. Cooke, H.E. Bez, "Computer Mathematics", Cambridge Computer Science Texts .18,1984,pg 339-365

Figure 1.

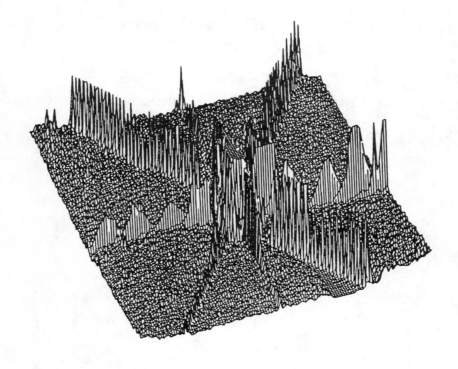

Figure 2.

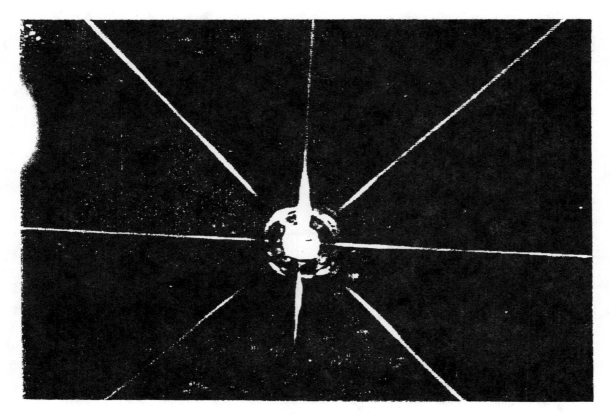

Figure 3.

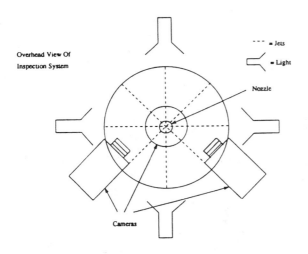

Figure 4.

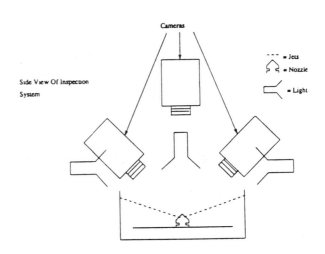

Figure 5.

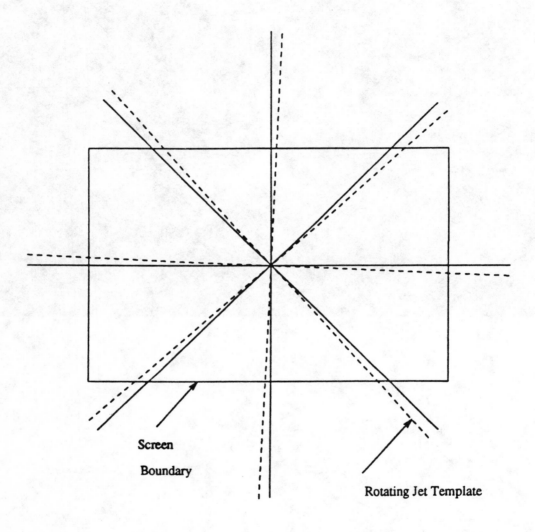

Screen

Boundary

Rotating Jet Template

Figure 6. Rotation Of Template To Find Orientation Of Sprays

End points	Coordinates (x,y,z)
point 0	$(\cos 18, 0, \sin 18)$
point 1	$(\cos 18 * \cos 45, \cos 18 * \sin 45, \sin 18)$
point 2	$(0, \cos 18, \sin 18)$
point 3	$(\cos 18 * \cos 135, \cos 18 * \sin 135, \sin 18)$
point 4	$(-\cos 18, 0, \sin 18)$
point 5	$(\cos 18 * \cos 225, \cos 18 * \sin 225, \sin 18)$
point 6	$(0, -\cos 18, \sin 18)$
point 7	$(\cos 18 * \cos 315, \cos 18 * \sin 315, \sin 18)$

Figure 7.

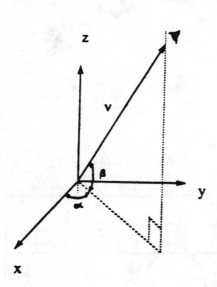

Figure 8

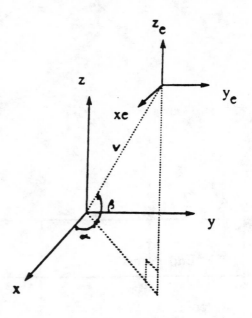

Figure 9

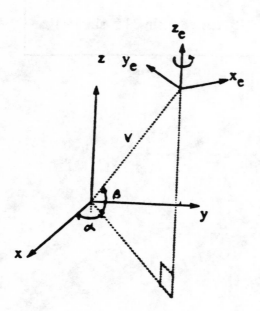

Figure 10

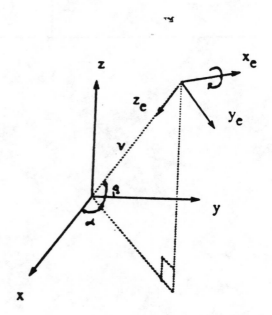

Figure 11

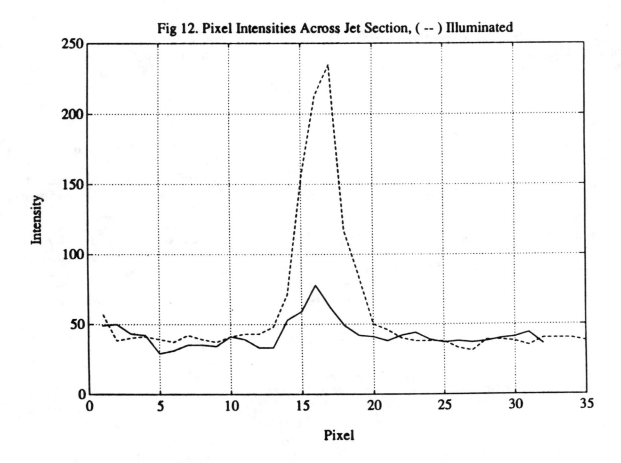

Fig 12. Pixel Intensities Across Jet Section, (--) Illuminated

A Scheme for the Analysis of Infinite Fringe Systems

J.C. Hunter, M.W. Collins & B.A. Tozer

Thermo-Fluids Engineering Research Centre, The City University, London, England.

ABSTRACT

Based on the practical experience gained during the development of an automatic fringe tracing and ordering programme at the City University Thermo-Fluids Laboratory, the paper presents a critical appraisal of available fringe analysis techniques, ending with a flow chart which delineates the necessary steps, and, where applicable, the viable alternatives for the one- or two-dimensional analysis of infinite fringe data.

One dimensional fringe analysis can be applied where only single scan data is required and the fringe analysis routines can be run essentially normal to the fringes. In this case, a choice of five viable techniques is available for the analysis, but automatic fringe ordering is impossible without extensive a priori knowledge, because of the lack of whole field data.

Two dimensional fringe analysis can be performed either by edge detection or by medial line extraction; the latter of these is time consuming but somewhat more accurate. Fringe ordering by manual, semi- or fully-automatic techniques is possible in this case. If only limited data storage capacity is available then it may be necessary to use a redundancy algorithm to reduce the amount of data to more moderate proportions.

1. INTRODUCTION

Interferometric fringe tracking routines vary according to the application to which they are directed. Thus the approach to attaining optimum performance will vary according to the nature of the fringe field - whether it is one dimensional (or approximately circularly symmetric), or two dimensional, and on whether it is necessary to measure small fractions of a fringe, or if simple fringe counting will suffice. The work described in this, and in the companion paper Hunter et al.[1], was undertaken to provide fringe tracking routines which are especially suited to problems in heat transfer and fluid flow, both one- and two-dimensional fringe fields being considered. The fringe fields obtained in these studies, taken with an infinite fringe background, typically contain many fringes and there is no call for measurements to small fractions of a fringe such as can be obtained using heterodyne and quasi-heterodyne techniques. In these the intensity of illumination at a point $\underline{u}$ in the field is given by:-

$$T(\underline{u}) = a_1{}^2 + a_2{}^2 + 2a_1 a_2 \cos (\omega_2(\underline{u}) - \omega_1 (\underline{u})) \tag{1}$$

where a_1, a_2 are the intensities due to the two successive interfering wavefronts and ω_1, ω_2 are the phases of these components.

2. FRINGE ANALYSIS TECHNIQUES FOR HOLOGRAPHIC INTERFEROGRAMS

2.1. General

All fringe analysis programmes involve the use of both image enhancement and fringe detection/tracking routines in various forms. A companion paper[1] discusses the use of image enhancement in fringe tracking problems and this paper concentrates on the tracking routines which can be used.

Fringe tracking routines may be divided into three main categories:-

a) Machine aided manual tracking routines.
b) One dimensional detection routines.
c) Two dimensional detection/tracking routines.

The machine aided manual tracking routines are somewhat outside the scope of this paper. A review of these routines can be found in Hunter[2].

2.2. One-dimensional Fringe Analysis Routines

2.2.1. <u>General</u>. These routines operate on one-dimensional grey-level signal data obtained either from a single scan line across the grey-level fringe field, or from a temporal or spatial averaging of a number of scan lines (to improve signal quality). To be effective the direction of the scan line has to be essentially perpendicular to the fringe direction. Five distinct techniques may then be applied for the digital determination of fringe co-ordinates:-

i) Locating the maximum and minimum grey-level intensities in the (assumed sinusoidal) signal.

ii) Determining regions of change in the grey-level gradients.

iii) Determining the regions of maximum grey-level gradient.

iv) Using a floating threshold or hysteresis technique to determine the positions of fringe maxima or minima.

v) Using a bucket and bin type algorithm to determine the positions of the fringe maxima or minima.

A schematic description of these five techniques is given in Fig.1.

2.2.2. <u>Fringe Detection Techniques.</u>

i) Max/Min fringe detection,[3,4,5]

With this method grey-level maxima and minima are identified and used as the fringe co-ordinates. Generally some threshold value, usually the average grey-level value over the scan line is identified, and maxima must lie above this threshold, minima below it.

This technique fails rapidly when signal-to-noise ratios fall below a relatively high figure. The technique is not recommended for general use.

ii) Grey-level gradient change detection,[6,7,8]

In this case fringe extrema are located along the scan line by detecting parts at which the grey level gradient over a short vector changes sign with respect to the previous short vector.

iii) Maximum gradient detection,[2]

This technique can be performed either by searching for a grey scale gradient that is greater than the two neighbouring gradients, or by employing a gradient sensitive operator and looking for a maximal response.

iv) Floating threshold fringe detection, [9,10,14]

This algorithm searches for the lowest (grey scale) valued pixel since the last maximum was detected. When found this pixel is assigned as "MIN". Then, if a pixel is found with a grey-level which exceeds "MIN" by a threshold T, the value and location of "MIN" are confirmed as a minimum. T is defined by:-

$$T = \Delta(A_r + A_s)/2 - ["MIN"] \tag{2}$$

where Δ is a constant (in the region 0.2 to 0.35), A_r is the average grey-level over some reference area (often the whole fringe system) and A_s is the average grey-level of the scan line. This type of threshold is used because, whilst it adjusts to varying impage brightness, it does not allow the threshold to approach zero in a predominantly dark region of the interferogram.

v) Bucket-bin fringe detection,[11]

A grey-level threshold is set, usually the mean grey-level encountered across the scan line, and from this various zero-crossing points are determined, e.g. PQ in Fig.2. All the points along PHQ then define a bucket, and the maximally valued pixel in the bucket defines the bin, which may be used to indicate the extrema. Alternatively, the bin pixel may be found by fitting a curve to the bucket pixels or by using a centre of gravity approach.

This basic algorithm may be improved upon the addition of a secondary mini-bucket, CHD in Fig.2, all subsequent data processing being performed within this mini-bucket.

2.2.3.   Fringe Validation.   The methods described above can be used to make an initial estimate of fringe co-ordinates, but some criteria are needed to validate the detected locations. Although to some extent these are application dependent, a typical series of checks might be:

a)   A maximum must follow a minimum and so on, alternately.

b)   The maximum intensities located must be at least 50% of the previously located maximum, and similarly the intensity of a minimum must be less than twice that of the previously located minimum.

c)   Some fringe spacing criteria might also be brought in, such as that all detected extrema should be at least one half, and not more than twice the previous semi-period apart.

Better fringe co-ordinate accuracy may also be obtained by using a least squares quadratic filtering routine, or by fitting a sinusoidal function and performing a correlative fitting procedure to the data points immediately surrounding the extrema.

2.2.4. Analysis of Broad Diffuse Fringes. When the fringes are broad and diffuse, a different approach has to be adopted. Schemm and Vest[12] have developed a procedure for smoothing and interpolating fringe position data when the number of fringes is small and there are more than 10 pixels per fringe. This consists of using a least squares fit criterion to fit the grey-level data over a single scan to a function of the form:-

$$I(x) = A(x) + B(x) \cos \Phi(x) \tag{3}$$

However for fringe fields of this character it is probably better, if at all possible, to use one of the direct phase measuring techniques, which are not reviewed here.

2.3. Two Dimensional Fringe Analysis Routines

2.3.1. General

One dimensional fringe analysis programmes are of only limited applicability and many fringe fields have a far too complex form for them to be handled in this way. In particular, as we shall see below, fringe ordering using one-dimensional methods is not possible without considerable prior knowledge, and then only in relatively simple cases.   It has therefore been necessary to develop two-dimensional fringe tracking routines.

The one-dimensional techniques reviewed above may be extended to two-dimensions either by analysing a series of scan lines and combining the results or by determining (say) the location of the bright fringe centres along a particular scan and then using a fringe tracking algorithm to trace these fringes.   These fringe tracking algorithms all operate by summing the grey-level intensities over a series of short vectors and (in the case of bright fringes) moving the centre point of the neighbourhood in the direction of maximum summed intensity.

A number of techniques exist for performing a two-dimensional analysis, the following having been considered here:

   i)   Segmentation.
   ii)   Fringe edge detection and polygonal data storage.
   iii) Feature Extraction algorithms.
   iv)   Redundancy reduction algorithms.
   v)   Fringe thinning, both serial and parallel processing.

These processes are applied after the raw data has been treated to image pre-processing and binarisation, both of which are considered in the companion paper[1].

2.3.2. Segmentation[13]. In an already binarised fringe field[1], fringes are taken to be sets of connected pixels of the same colour, either black or white. The procedure starts in the top left hand corner of the image data, using the first pixel as the seed pixel. The algorithm then tests the four neighbours of each pixel connected to this seed pixel, and each neighbour of the same colour is labelled "A", and used recursively as a new seed pixel. After the first fringe has been labelled "A" the image is scanned to find a pixel that has not been labelled, and this pixel is labelled "B" and becomes the initial seed pixel for the next fringe, and so on until all the fringes have been segmented, Fig.3.

Fringe edges are detected by locating regions where pixel labelling changes, and using a boundary fill routine to trace these changes.

2.3.3. <u>Fringe Edge Detection and Polygonal Data Storage</u>. Becker and Yu[14] described a system in which the fringe co-ordinates are determined at the transition from black to white and vice versa. A sequential tracking procedure was then used to trace the fringe positions line by line and the fringe field stored as a polygonal data structure.

2.3.4. <u>Feature Extraction Algorithms.</u>[10,15]. In such an algorithm the left and right edge points of each fringe encountered along the scan line are compared to the left and right edge points derived from the next scan line. Each of the edge points in the current scan line is assigned to a fringe polygon by a simple boundary continuation test, the action to be taken depending on the determined continuation conditions, see Fig.4.

2.3.5. <u>Fringe Data Compression</u>. An important requirement when large amounts of fringe data are to be processed is the effective storage of such data. A useful scheme is the approximation of the fringe curve by polygons[16], with a minimum number of vertices and satisfying a given fit criterion, Fig.5. This is really an extension of the Freeman's chain code idea, in that the number of polygonal vertices are minimised.

2.3.6. <u>Fringe Thinning Algorithms.</u> These are the most popular method for determining fringe co-ordinate data, despite the fact that they require extensive computation and are therefore time-consuming (see Fig.6). Thinning algorithms work by eroding the boundaries of a feature until only a fully connected medial line remains. This can be done by parallel or by serial processing.

A typical example of a parallel processing algorithm[17] operates as follows:

Let $B(p_1)$ be the number of non-zero neighbours of p_1. Then a point p_1 is deleted from the figure if:

(i)	$2 \leqslant B(p_1) \leqslant 6$	and
(ii)	$A(p_1) = 1$	and
(iii)	$p_1, p_3, p_5 = 0$ or $A(p_2) \neq 1$	and
(iv)	$p_2, p_4, p_6 = 0$ or $A(p_4) \neq 1$.	

This algorithm yields corrected skeletons which are not susceptible to contour noise.

Seguchi et al.[18] used a parallel processing thinning algorithm to extract fringe contours from binarised fringe data. To group the data into fringe sets the image is scanned in a raster, from top left to bottom right and when a fringe contour is located it is traced using a simple neighbourhood search scheme, assuming both total connectivity of the contour and that the contour is only one pixel wide.

In serial processing, one point at a time is processed. The result of processing a point at the z_{th} iteration depends on a set of points for some of which the result of the zth iteration is already known. Although serial processing can be more efficient than parallel processing when it is implemented on a general purpose computer, it does not generally work as well as parallel processing algorithms. Suitable algorithms have been described in the literature[17,19,20].

2.4. Fringe Ordering Schemes

2.4.1. Fringe Ordering with One-dimensional Routines.
Because no full field information is available in these circumstances there are severe limitations on automatic processes. Ordering is usually performed manually, with some form of interactive device such as a light pen or a cursor. Ordering can only be performed automatically in this case if three conditions are met:

a)   The initial fringe order is constant or is manually set.

b)   An a priori knowledge of the probable fringe order distribution can be built into the algorithm (e.g. "the fringe order will always increase").

c)   There must be some form of interactive facility available so that mis-ordered fringe locations can be corrected.

2.4.2. Fringe Ordering with Two-dimensional Routines.
Although in most cases fringe ordering is done manually, the two-dimensional fringe analysis routine gathers sufficient information for automatic ordering schemes to be introduced in most cases and Becker et al.[10] described two.

If no discontinuities exist in the fringe field a simple semi-automatic numbering scheme is applicable. This involves dividing the fringe field into regions containing similarly behaved fringes, connecting these by a straight line, and numbering sequentially along the intersections, starting from the greatest fringe density. Some a priori knowledge is required to obtain absolute fringe orders.

In the many cases where discontinuities occur a different approach must be adopted. The fringe system is windowed into (typically) 64x64 pixel regions and fringes numbered window by window starting with the one having the maximum number of fringes combined with a minimum of inversion points. An adjacent window is then searched which has at least two (already numbered) fringes at the common grid line and which also conforms to the initial criteria. In this way meshes containing discontinuities are processed last, when a great deal is already known about the fringe orders in the system.

2.5. Summary

The infinite fringe analysis techniques are summarised in the algorithm, Fig.6.

3. THE CITY UNIVERSITY FRINGE ANALYSIS PROGRAMME

3.1. General

The fringe analysis programmes developed at the City University for the analysis of heat transfer and fluid flow interferograms have been described elsewhere[2,21,22].

Programmes were developed for both one- and two-dimensional fringe analysis procedures and the following is a brief summary of the techniques employed.

3.2. One-dimensional Fringe Analysis

3.2.1. Section 2.2.2 above describes five ways of utilising the grey scale data to extract fringe co-ordinates. All five were tested using computer generated fringe data in which the positions of the fringe maxima were known, but the data presented for analysis could be corrupted in a controlled manner by varying the signal-to-noise ratio and the fringe visibility. Using this test programme it was possible to compare the effectiveness of each programme in terms of ultimate fringe detectability and also to compare the processing time required. Furthermore, the basic algorithms used for each fringe detection technique could be developed and optimised against known criteria. Details of the algorithms finally employed are given in Hunter[2].

3.2.2. <u>Results of Fringe Detection Tests</u>. a) The effects of a varying fringe visibility on a noise free image were tested and the results are shown in Fig.7. These show that the bucket-bin and grey-level gradient methods perform best for this test.
b) The effects of a varying signal-to-noise ratio were next tested for and results are shown in Fig.8. (Note that the axes in this figure are inverted from those in Fig.7). On this test the gradient change and bucket-bin algorithms proved to be the most effective, the maximum gradient method being less satisfactory.

These results showed that the gradient change and bucket-bin routines were most effective for one-dimensional fringe co-ordinate location and these were adopted for the programme. A further test on the effect of varying gradient length on the gradient change algorithm showed that it is important to keep the length over which gradient is measured to less than about 0.75 x fringe width, when possible.

3.3. Two-dimensional Fringe Analysis

3.3.1. <u>Fringe Co-ordinate Extraction</u>. A sequential thinning algorithm due to Rosenfeld and Pfaltz[19] was tested but was found to work poorly on the proferred data. Parallel thinning algorithms due to Hilditch[17], Arcelli et al.[23] and West[24] were tried and it was found that (a) they were computationally expensive, (b) they tended to produce false features such as small sections of false medial lines branching from true medial lines, (c) erroneous fringe information results if the field segmentation has been poorly performed.

For these reasons a technique involving fringe edge detection was evolved. The advantages of this technique which is fully described in Hunter[2] are as follows:

1. The algorithm operates rapidly, taking approximately the same time as a single pass of the iterative fringe thinning routines.

2. Two sets of ordinates are produced for each fringe position as opposed to one medial line obtained by fringe thinning.

3. Since fringe edges define complete contours, any point where branching occurs can be examined to determine which branch was a false structure.

4. As the fringe data itself is not altered it is possible to correct edge data for regions which have been wrongly segmented.

3.3.2. <u>Fringe Tracking and Ordering.</u>   The fringe tracking routine followed conventional methods and is described elsewhere[2].   At first a fully automatic fringe ordering routine was tried, but this proved to be unreliable when used with the kinds of fringe fields of interest in this work.   In particular it was found to be unable to cope with shock regions, giving false orders to fringes in surrounding regions.   In addition, closed fringe systems could not be adequately determined by the programme, it being necessary to have a priori knowledge in order to determine whether fringe orders increased or decreased.   A semi-automatic fringe ordering system was therefore adopted.

3.3.3. <u>Operation of the Programme.</u>   Fig.9 shows a flow diagram of the programme operation.   Fig.10 shows a photograph from a complete holographic record of a flow interferogram, and Fig.11 shows a section of this field after processing, tracking and ordering using this programme.   The time taken was about 10 minutes.

4.   CONCLUSIONS

Although fringe analysis systems based on direct phase measurement techniques have much to offer in terms of complete computational automation and high measurement accuracy, there remains a place for the use of fringe tracking techniques in many applications where high fringe densities occur, as well as in non-interferometric applications.   In these cases, it is possible to save a great deal of time and effort by applying the optimum image enhancement, fringe reduction and location and fringe tracking routines, but it is still often not possible to adopt fully automatic fringe ordering for all applications.

5.   REFERENCES

1.   J.C. Hunter, M.W. Collins & B.A. Tozer, *Procs. SPIE Conference,* San Diego, Aug.1989.

2.   J.C. Hunter, Ph.D. Thesis.   City University, London, 1987.

3.   P. Varman, "A Moiré System for Producing Numerical Data of the Profile of a Turbine Blade Using a Computer and Video Store".   Optics & Lasers in Engineering, <u>5</u>, pp.41-58, 1984.

4.   W.R.J. Funnel, "Image Processing Applied to the Interactive Analysis of Interferometric Fringes".   Appl.Opt., <u>20</u>, p.3245, 1981.

5.   T. Yatagi, S. Inaba, H. Nakaro, & M. Suzuki,   "Automatic Flatness Tester for Very Large Scale Integrated Circuit Wafers".   Opt.Eng., <u>23</u>, p.401, 1984.

6.   K. Birch,   "A TV Based System for Interferogram Analysis".   SPIE, <u>369</u>, p.186, 1976.

7.   D.W. Robinson,   "Automatic Fringe Analysis with a Computer Image Processing System".   Appl.Opt., <u>22</u>, p.2169, 1983.

8.   D.W. Robinson,   "Role for Automatic Fringe Analysis in Optical Metrology".   SPIE, <u>376</u>, p.20, 1983.

9. D.A. Tichenor & V.P. Masden, "Computer Analysis of Holographic Interferograms for Non-Destructive Testing". Opt.Eng., 8, p.469, 1979.

10. F. Becker, G. Meier, & H. Wegner, "Automatic Evaluation of Interferograms". SPIE, 359, 1982.

11. A. Choudry, "Digital Holographic Interferometry of Convective Heat Transport". Appl.Opt., 20, 7, p.1260, 1981.

12. J.B. Schemm & C.M. Vest, "Fringe Pattern Recognition and Interpolation using Non-linear Regression Analysis". Appl.Opt., 22, 18, p.2850, 1983.

13. H. Cline, A. Holik, & W. Lorenson, "Computer-Aided Surface Reconstruction of Interference Contours". Appl.Opt., 21, p.4481, 1982.

14. F. Becker & Y. Yu, "Digital Fringe Reduction Technique Applied to the Measurement of Three-dimensional Transonic Flow Fields". Opt.Eng., 24, 429, 1985.

15. A. Agrawala & A. Kulkarni, "A Sequential Approach to the Extraction of Shape Features". Computer Graphics and Image Processing, 6, pp.538-557, 1972.

16. U. Ramer, "An Iterative Procedure for the Polygonal Approximation of Plane Curves". Computer Graphics and Image Processing, 1, pp.244-256, 1972.

17. C.J. Hilditch, "Linear Skeletons from Square Cupboards". Machine Intelligence, Vol.iv, Eds.B.Meltzer and D.Mitchie, Elsevier, N.Y. pp.403-420, 1969.

18. Y. Seguchi, Y. Tomita, & W. Watanabe, "Computer-Aided Fringe Pattern Analyser - A Case of Photoelastic Fringe". Expt.Mech., 19, p.362, 1979.

19. A. Rosenfeld & J. Pfaltz, "Sequential Operations in Digital Picture Processing". J.A.C.M., 13, 4, pp.471-494, 1966.

20. U. Montanati, "A Method for Obtaining Skeletons using a Quasi-Euclidean Distance". J.A.C.M., 15, pp.600-624, 1968.

21. J.C. Hunter & M.W. Collins, "Problems in using Holographic Interferometry to Resolve the Four-dimensional Character of Turbulence, Part II: Image and Data Processing". J. Opt. Sensors, 1, 3, pp. 1986

22. J.C. Hunter & M.W. Collins, "Holographic Interferometry and Digital Fringe Processing". J.Phys.D., 20, pp.683-691, 1987.

23. C. Arcelli, L. Cordella, & S.Levialdi, "Parallel Thinning of Binary Pictures". Electronics Letters, 11, No.7, p.148, 1975.

24. G.A.W. West, Private Communication, 1986.

Fig.1

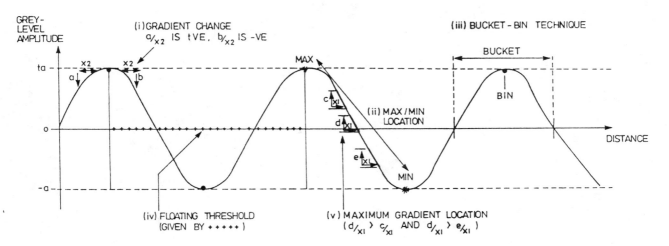

GREY-LEVEL AMPLITUDE

(i) GRADIENT CHANGE
$a_{/x_2}$ IS +VE, $b_{/x_2}$ IS -VE

(iii) BUCKET - BIN TECHNIQUE

BUCKET

BIN

MAX

ta

x_2 x_2

a b

$c_{/x_1}$

(ii) MAX / MIN LOCATION

$d_{/x_1}$

0

$e_{/x_1}$

DISTANCE

MIN

-a

(iv) FLOATING THRESHOLD
(GIVEN BY + + + +)

(v) MAXIMUM GRADIENT LOCATION
($d_{/x_1} > c_{/x_1}$ AND $d_{/x_1} > e_{/x_1}$)

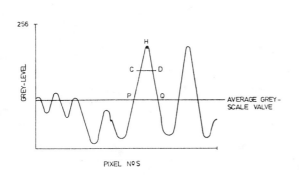

256

GREY-LEVEL

H

C   D

P   Q

AVERAGE GREY-SCALE VALVE

PIXEL NºS

Fig.2

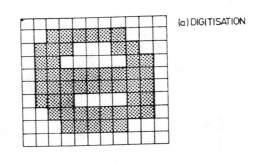

(a) DIGITISATION

Fig.3

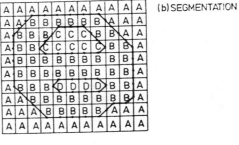

(b) SEGMENTATION

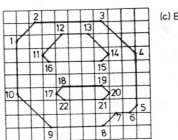

(c) EDGE DETECTION

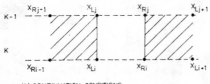

K-1

K

(i) CONTINUATION CONDITIONS

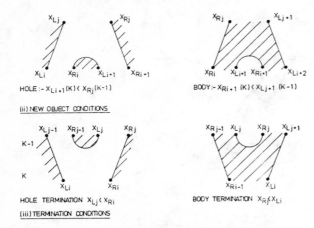

HOLE :- $X_{Li+1}(K) < X_{Rj}(K-1)$

BODY :- $X_{Ri+1}(K) < X_{Lj+1}(K-1)$

(ii) NEW OBJECT CONDITIONS

K-1

K

HOLE TERMINATION $X_{Lj} < X_{Ri}$

BODY TERMINATION $X_{Rj} < X_{Li}$

(iii) TERMINATION CONDITIONS

Fig.4

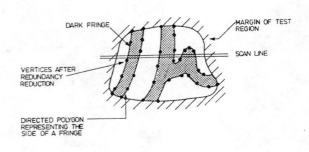

Fig.5

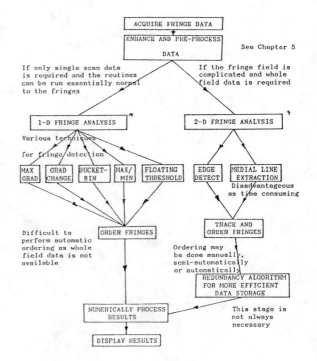

Fig.6

Fig.7

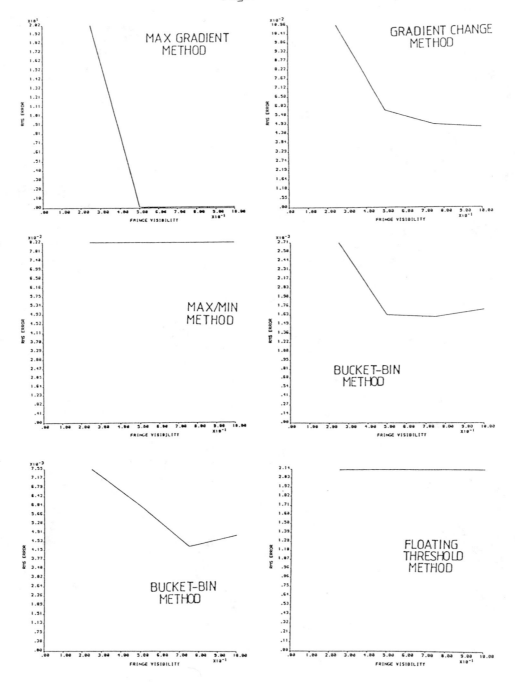

Fig.8

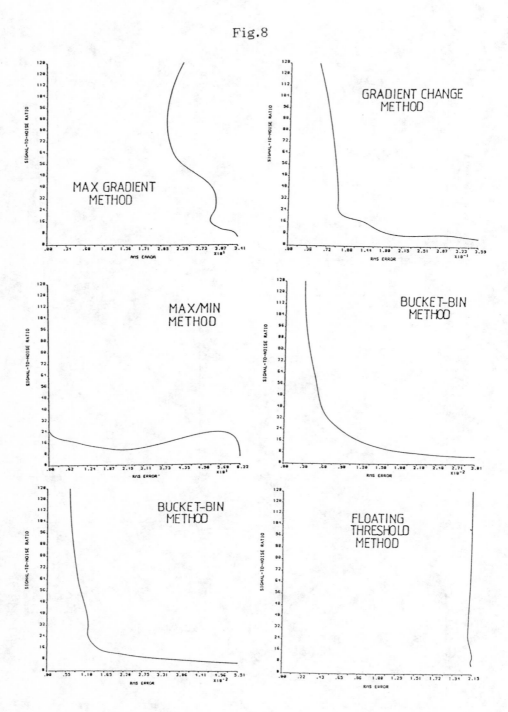

Fig.9.

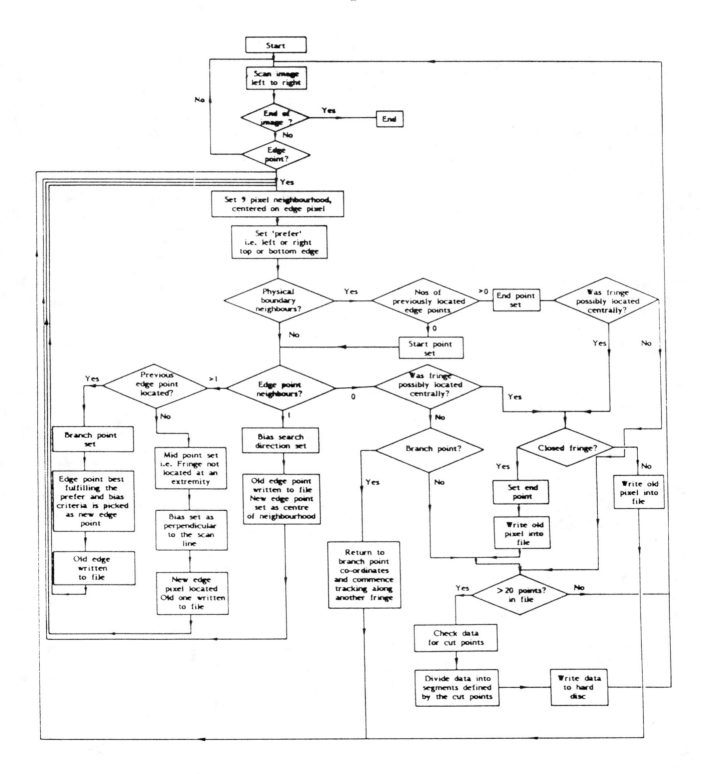

Spectral contents analysis of birefringent sensors

Alex S. Redner

Strainoptic Technologies, Inc.
108 West Montgomery Avenue, North Wales, Pennsylvania 19454

1. INTRODUCTION

Fiber-optic sensors are becoming increasingly popular. Their use is steadily expanding and a large variety of fiber-optic transducers were designed and are commercialized for measuring pressure, strain, temperature, etc.

The electric-resistance bonded strain-gage transducer uses a spring element to convert the measured quantity into resistance changes and provides a directly measurable passive output.

The fiber-optic sensor by contrast uses a variety of sensing concepts, including micro-bends, tip displacement, radiation, interferometry, and birefringence.

The interferometric and birefringent sensors are of particular interest because of their high-precision potential. In both instances, the same fundamental physical event takes place: the beam of light is divided into two paths and a relative path-difference is created, as a result of the event, ε that is measured. The two beams are re-combined and allowed to interfere.

In an interferometric sensor, the difference in the path length between the "reference" and "active" fibers is proportional to the measured event ε. (see figure 1a).

$$\delta = K \varepsilon \qquad (1)$$

where δ is the relative retardation
ε is the path-length change in the active beam, proportional to the measured event.
K a coefficient of proportionality

In a birefringent sensor, both the "reference" and "active" rays cross the sensor together, but remain polarized in mutually perpendicular planes. (see figure 1b), and the relative retardation between the two rays becomes:

$$\delta = t \times C_B \times \varepsilon \qquad (2)$$

where t is the sensor path length
C_B Brewster constant
ε measured event converted into the sensor stress

In both instances, the interference of "active" and "reference" beams yields similar light intensity relation:

Interferometric Fringes $I = I_0 \cos^2 \dfrac{\pi \delta}{\lambda}$

Photoelastic Fringes (birefringent) $I = I_0 \cos^2 \dfrac{\pi \delta}{\lambda}$ (light field)

or $I = I_0 \sin^2 \dfrac{\pi \delta}{\lambda}$ (dark field)

$$(3)$$

In monochromatic light, photoelastic and interferometer fringes look alike and have the same meaning. They are 1 oci of the constant value of δ (Figure 2). Fringe orders can be assigned as integers, where $\delta = 0, 1\lambda, 2\lambda, \ldots n\lambda$, etc., but the absolute order of the observed fringe n remains unknown.

$$\delta = \frac{\lambda}{\pi} \text{ arc sin } \left(I/I_0\right)^{\frac{1}{2}} \pm n\lambda \tag{4}$$

The light intensity I is related to the retardation δ. Using light intensity measurements, the retardation δ can be measured at a point using fiber-optic transmission. Since the integer n is not known, one can limit the retardation δ to $\frac{1}{2}\lambda$, thus creating a well-defined relationship between the measured light intensity and the event $\mathcal{E}$.

This concept can be implemented in a birefringent sensor, but is limited to those instances where the maximum value of $\mathcal{E}$ is well defined. Another solution is to allow the path length δ to become very large, and count the fringes. This approach requires a positive identification of the fringe count sign, and adds to the complexity of the system.

In the recently developed new method (1), (2), (3), (4), the monochromatic light is replaced by white light.

The equation above (1), (2), (3) remains essentially unchanged. The interferometric and photoelastic fringes are observed in white light as a set of isochromatic (color) fringes well known in photoelasticity and also frequently observed in interferometry as "Newton rings" and "air-gap" fringes. The nature of these colors can be easily interpreted using the figure 3, where the light intensity is traced vs. wavelength λ, for various retardation levels δ. In the figure 3, a stressed member is placed in a dark-field polariscope and the theoretical light intensity (equation 3) is traced vs. wavelength in the visible region. The observed sequence of color vs. retardation is readily seen, but a visual observation cannot assign the value of δ, or distinguish between the first second or third red or green.

2. MEASURING RETARDATION USING SPECTRAL CONTENTS ANALYSIS (SCA) (5)

The SCA method, developed for accurately measuring δ, uses a set-up schematically shown in figure 4.

After crossing an interferometric or photoelastic sensor, the light is analyzed using a dedicated spectrophotometer, that measures the light intensity at several wavelengths λ_j.

Introducing a system factor S_λ, that includes the spectral distribution of the source, chromatic variation of absorbence and losses occuring in all optical components placed in the path of the beam, we can rewrite the equation 3 as follows:

$$I_\lambda = S_\lambda \sin^2 \frac{\pi\delta}{\lambda} = S_\lambda \times T_M (\delta, \lambda) \tag{5}$$

In this equation, I_λ is the light intensity at wavelength λ, and $T_M (\delta, \lambda_j)$ is the spectral "transmittance" of the birefringent (or interferometric) sensor. For any measured event $\mathcal{E}$, there is only one value of δ (defined by equations 1 and 2) and therefore, a unique "spectral" signature $T_M(\delta, \lambda)$, as illustrated by the example shown in figure 3.

In the SCA method, the light intensity $I_{\lambda j}$ is measured at several wavelengths, using a photodiode array, with each photodiode measuring in a narrow interval $\Delta \lambda$. Using m-photodiodes array, m-data points are measured to identify δ from m equations:

$$\sin^2 \frac{\pi \delta}{\lambda_j} = \frac{I_{\lambda j}}{S_{\lambda j}} \quad (1 \le j \le m) \tag{6}$$

This system of equation is numerically solved using a PC-AT based data acquisition system, shown schematically in the Figure 4.

The concept of the SCA system was demonstrated (3) in a feasibility study. The system was further refined to permit multipoint data acquisition remotely, and to dynamically acquire the data.

3. BIREFRINGENT SENSOR FOR ELEVATED TEMPERATURE

While in principle the SCA method can be used with both interferometric and birefringent sensors, in practice the birefringent sensor offers several significant advantages.

A temperature change introduces thermal expansion in all sensors (interferometric, strain gage, ...) and requires a carefully designed compensation leg, subjected to the same temperature as the active leg. This is not usually possible in a high-temperature-gradient environment.

The photoelastic response of a birefringent sensor is proportional to the difference of principal strains:

$$\delta = K \, t \, (\varepsilon_1 - \varepsilon_2)$$

Consequently, the birefringent sensor does not respond to temperature changes, and it does not require a zero-shift compensation.

The range of the birefringent sensor is unlimited, since there is a broad choice of control parameters, including the path t, strain level, etc.

The choice of sensor material is crucial, since it defines the temperature limitations of the transducer. The Table I illustrates materials that were considered for use. In our experiments, fused silica was selected, since it satisfied the temperature range in the sensor evaluation program, was readily available, and was procurable at a reasonable cost in a variety of geometries.

4. POLARIZERS

The polarization of light entering and emerging from the sensor, operating at elevated temperature, can be accomplished using a polarization-preserving fiber. At the time of testing reported here, polarization-preserving fibers, capable of sustaining 2000° F, were not available. Instead, Brewster angle plates were used. The availability of high-temperature polarization-preserving fibers will make the implementation of the birefringent sensor considerably easier.

5. MATERIAL CALIBRATION

Photo figure 5 illustrates the experimental set-up used to measure the Brewster constant of fused silica, and also to evaluate the SCA method of data acquisition.

Calibration of fused silica was carried on a $1\frac{1}{2}$" diameter disc subjected to diametrical compression (figure 6). The force was measured using a calibrated load cell. The birefringence was measured both visually, then using the SCA method at 200° C, 422° C, 730° C and 1100° C and the Brewster constant was found to be constant up to this temperature.

6. ADVANTAGES A LIMITATIONS OF SCA

The SCA is easily applicable to the data acquisition for a birefringent sensor. The range of the measured retardation is limited to 10μ or less, and the resolution is dictated by the required speed of response. Using an error function (1), (3) for data analysis, a resolution 2 nm was easily accomplished (.02% of the range).

The advantages of the SCA method over a monochromatic system are numerous. The system is yielding retardation statiscally by curve fitting to a broad sample of measured light intensities and, therefore, benefits from its statiscal aspect. The calibration and the absolute precision depends on the spectrophotometer calibration only and is unaffected by the environment of the sensor. The light losses have virtually no effect on the measurements, since only the "signature" is identified.

The SCA system prototype was constructed under NASA Phase II SBIR program and evaluated. Following the prototype development, a commercial system, developed by Strainoptic Technologies, Inc., is presently used in industrial applications for ON-LINE monitoring of strains in glass and orientation of polymers and films.

7. REFERENCES

1. Redner, A.S. "Photoelastic Measurements by Means of Computer-Assisted Spectral Contents Analysis", _Proceedings_ of the 5th International Congress on Experimental Mechanics, Montreal, Canada, 1984, pp. 421-427.

2. Sanford, R.J. "On the Range and Accuracy of Spectrally Scanned White Light Photoelasticity", _Proceedings_ 1986 Spring Conference on Experimental Mechanics, New Orleans, Louisiana, 1986, pp. 901-908.

3. Phase I report, contract #NAS2-12351.

4. U.S. Patent #4,668,086, issued 1987.

8. ACKNOWLEDGEMENTS

The feasibility study of the SCA method application to measure strains at elevated temperature and the development of the SCA system were supported by NASA SBIR R&D Phase I and II contract #NAS2-12666. This support is gratefully acknowledged.

Mr. Alan Carter of NASA-Ames Research Center provided many helpful suggestions and and enthusiastic support. His contribution was instrumental to the successful development of this new method.

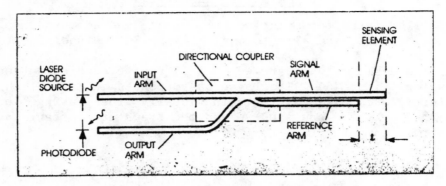

Figure 1a. Interferometric sensor

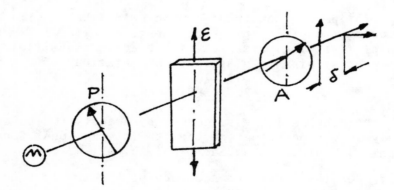

Figure 1b. Birefringent sensor

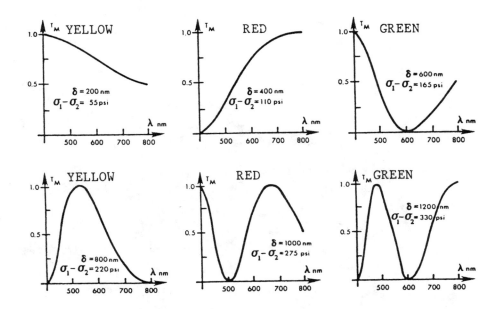

Figure 3. Spectral signature at increasing retardation levels

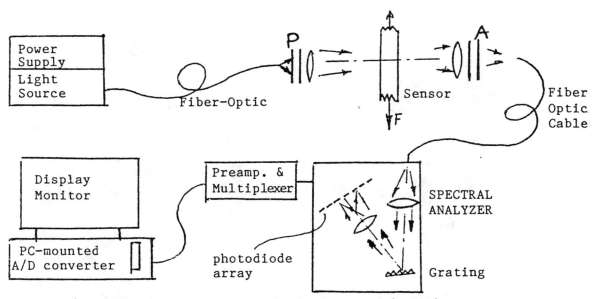

Figure 4. Spectral Contents Analysis System Schematic

Figure 5.   Experimental Set-Up for Measuring Brewster Constant at
Elevated Temperature.

Figure 6.   Fused Silica Disk and Compression with Loading Fixture.

TABLE I. BIREFRINGENT-SENSOR MATERIALS FOR ELEVATED TEMPERATURE

Material	Formula	Melting Point	Crystaline Structure	Index of Refraction
Aluminum Nitride	AlN	> 2200	hexagonal	
Sapphire	Al_2O_3	2053	rhombohedral	1.765
	Be_3N_2	2200	cubic	
	CaO	2580	cubic	1.838
	$CaZrO_3$	2550	monoclinic	
	CeO_2	2600	cubic	
Spinel	$MgAl_2O_4$	2135	cubic	1.723
Periclase	MgO	2800	cubic	1.736
Silicon Carbide	SiC	2600	hexagonal or cubic	2.645 2.697
Silica	SiO_2	1800	amorphous	1.456
Zirconia	ZrO_2	2700	monoclinic below 1000 C	
			Cubic above 1000 C	
Zircon	$ZrSiO_4$	2550	tetragonal	1.92-1.96 1.97-2.02

Correlation of Fringe Patterns using
Multiple Digital Signal Processors

D.A.Hartley, C.A.Hobson, S.Monaghan

Coherent and Electro-Optics Research Group
Liverpool Polytechnic U.K.

ABSTRACT

Image processing applications are computationally expensive. The
paper shows that the computational time required for an application
such as correlation can be considerably reduced by using a hierarchy
of images. Even with these optimisation techniques, speeds do not
approach real time on a uniprocessor system, but significant speed
increases can be obtained by using a multi-processor system. It is
shown that choosing the correct processor for the desired application
is an important decision and that for correlation a digital signal
processor provides superior performance over conventional processors.

1. INTRODUCTION

The use of image processing methods has become widespread over
recent years with applications ranging from medical diagnosis to
space exploration. Recent work at the Polytechnic has involved the
use of image processing methods and fringe analysis techniques to
determine the surface wear in dental restorations[1].

This work involved projecting fringe patterns onto the surface
under inspection. The resulting image was then digitised and stored
in a computer. After several months of wear, fringes were again
projected onto the surface and a second image obtained. A
comparative analysis of the difference between the two images was
then performed which produced information concerning the change in
surface form. This technique required that the second image be
situated in exactly the same location as the original image. As the
alignment and relocation was performed manually, this gave rise to
errors and was a time-consuming process. Computer automation has
been used to remove the alignment errors and to increase the speed of
operation.

2. CORRELATION METHODS

Correlation has been used in a large number of applications
including meteorology and radar systems. A simple measure of
association is given by the two dimensional form of the Correlation
Function[2] -

$$R(m,n) = \sum_x \sum_y f(x+m,y+n).W(x,y) \qquad (1)$$

where f(x+m,y+n) is part of the image at offset (m,n) and W(x,y) is the correlation template (the image to be matched). Large values of R indicate similarity between the image 'f' and the correlation template 'W', whilst small values indicate little or no similarity.

This equation will compute the correlation function at one position of the image. To search the entire image requires that the correlation function be calculated at each point. If the image is of size MxN, the correlation template is of size kxL and if the template is not allowed to move off the image then the number of correlations to be calculated is given by

$$[(M-k+1) \ x \ (N-L+1)] \tag{2}$$

Equation 1 was implemented in PL/M-86 on a 5MHz 8086/8087 based Intel System 86/380 and took eighteen hours to obtain the correct match for a template size of 128x120 pixels over an area of 128x128 pixels. This excessive time is caused by the large number of repetitive calculations which have to be performed - over 251 million multiplications and additions.

3. PYRAMIDAL CORRELATION

By using a hierarchical structure of images at various resolutions, the time required to perform correlation can be considerably reduced. Before correlation is applied, the correlation template and the image need to be reduced in resolution. This is achieved by averaging four pixels into one which is repeated until the desired resolution is obtained. This process is called consolidation. The software which performed the consolidation process produced four separate images with resolutions of - the original resolution, a quarter of the original resolution, a sixteenth of the original resolution, and finally a sixty fourth of the original resolution.

After consolidation, correlation was performed at the lowest resolution with the entire Level 4 image being covered. This produces a position of best match which is then used as the starting point at the next image level. The correlation function is then calculated only at the previous best match position and the eight neighbouring pixel locations. After the best match position has been found at this level the next level is used. This is repeated until the highest resolution is reached; the position of best match at this resolution is the overall best match. For example, if an image is of size M=120, N=128 and a template of size k=64, L=64 then, using equation 2, single level correlation requires that 3705 correlation function values be calculated whilst four level pyramidal correlation requires only 602 correlations to be calculated. This considerable reduction in the required computations reduced the program runtime to twenty minutes. This dramatic reduction in time indicates that the added complexity of pyramidal correlation is worthwhile.

4. REVIEW OF IMAGE PROCESSING HARDWARE

Increased performance can be obtained by running existing programs on faster computers. For example, the Intel System 86/380 could have its 5MHz 8086/8087 board replaced by a significantly faster 25MHz 80386/80387 board. Even by doing this the performance will not approach real time speeds due to the single processor sequentially working through the required computation. Image processing involves repetitive operations on the pixel elements - therefore to increase the speed of programmable systems parallelism may be used.

Parallel computer architecture can be divided into two main classifications[3]; these are Single Instruction Multiple Data (SIMD) and Multiple Instruction Multiple Data (MIMD).

SIMD systems are typically comprised of a single control unit (CU) which broadcasts instructions to an array of processing elements (PE). Each PE executes the same instruction. The PE is usually bit-serial in nature meaning that it operates on one bit at a time. This is not a disadvantage as the simplicity of a bit serial processor enables large numbers to be incorporated onto a VLSI chip. Examples of SIMD systems are the 1024 processor element AIS-5000[4] arranged in a linear array, the 16384 processor element MPP[5] arranged in a 128x128 array and the 65536 processor element Connection Machine[6] organised as four 16384 processor element arrays.

MIMD systems consist of processors which generate and execute their own instructions. Each processor can either have its own local memory with communication between processors being performed via message passing along communication links, or processors can share memory with interprocessor communication being carried out through the shared memory. Each processor is usually bit-parallel and can be either a conventional microprocessor or a custom VLSI processor. Examples of MIMD systems include the 8086 based Cosmic Cube[7] and the custom VLSI processor based NCube[8]. A processor specially designed for parallel processing is the Transputer and these processors have already been incorporated in image processing systems[9].

Image processing systems can also be constructed from dedicated hardware. Individual image processing operations are implemented in separate hardware circuits. With the advent of custom VLSI technology this has enabled high performance systems to be constructed using a small number of components. Examples of applications which have been implemented by dedicated hardware include Roberts product edge detection[10] and pattern recognition[11].

This brief review indicates that a wide variety of architectures are possible when designing an image processing system. If the application involves only a single task then dedicated hardware is the best solution. If, however, the application requires that the system be versatile then a programmable system is required. SIMD systems are more suited to applications such as edge detection, filtering and thresholding, whilst MIMD systems are suited to contour tracing, shape analysis and object classification. Therefore the application will determine which architecture is best.

5. CHOICE OF PROCESSOR

The decision to construct a MIMD system was based upon the requirement that the system be used for both low and high level tasks, the existing software would map easily onto a MIMD system, and it could be designed and constructed from readily available components.

The choice of processor is very important. The processor should be capable of 1) performing multiply and accumulate (MAC) operations efficiently and 2) contain a large address space to hold the image data. Modern microprocessors certainly meet requirement 2), however they do not efficiently implement MAC operations. Recently released RISC processors satisfy both of the above requirements and the Inmos T800 Transputer[12], which includes many RISC design philosophies, has been evaluated. A range of processors specially designed for MAC operations are Digital Signal Processors. These processors have an onboard hardware multiplier which is capable of performing the MAC operation in a single cycle. Digital Signal Processors are available from several manufacturers with the Texas Instruments devices having the largest installed user base. The Texas Instruments TMS320C25[13] has been evaluated.

6. PROCESSOR EVALUATION

Correlation software was written for both the T800 and the TMS320C25 to evaluate the performance of each device. The image and template size varied from a 3x3 pixel array up to 256x256 pixel array. For each image/template size two tests were performed. In the first test each 8-bit image and template pixel was simply stored in a separate memory location. Thus for the T800 each pixel was stored in a 32-bit word, whilst for the TMS320C25 each pixel was stored in a 16-bit word. This type of pixel storage has been called Unpacked Pixels and is shown in figure 1. This method of pixel storage provides the easiest access to pixels but consumes a large amount of memory. Memory requirements for image and template data can be reduced by placing several pixels into each 32-bit or 16-bit word. For the TMS320C25 two pixels are stored in each 16-bit word whilst for the T800 two pixels are stored in each 32-bit word. This method of pixel storage has been called Packed Pixels and is shown in figure 2. This storage method reduces memory requirements but involves an unpacking process to extract the relevant pixels.

7. IMPLEMENTATION DETAILS

As the T800 contains an onboard floating point unit, the correlation software could use this or the integer ALU to perform the calculation. Two programs were written to investigate the

performance of each maths unit. The software was written in the
OCCAM 2 language[14] and the programs run on a 20MHz T800 Transputer.

For the TMS320C25, the software was written in assembler and
executed on a TMS320C25 software simulator run on a conventional PC.

The unpacked pixel software simply involves multiplying the
relevant pixels and accumulating the products. As pixels are 8-bit
numbers a conversion to floating point format is required for the
T800 floating point correlation software. The packed pixel software
involves -

a) loading the packed pixel data into the processor
b) temporary saving this data
c) shifting right eight bits to obtain the lower order pixel
d) saving the low order pixel
e) recalling the temporary saved data
f) masking off the upper bits
e) saving the high order pixel

after this unpacking the individual pixels can be multiplied and
accumulated.

8.   ANALYSIS OF CORRELATION TIMES

The correlation times obtained were used to calculate the total
time required to perform pyramidal correlation. When using the T800
with unpacked pixel data the time to perform pyramidal correlation
was 28.56 seconds for integer arithmetic and 27.37 seconds for real
arithmetic. With packed pixel data, these times were reduced to 25.3
seconds and 25.77 seconds respectively. Using the TMS320C25 the
unpacked pixel time was 1.28 seconds whilst using packed pixels
increased the pyramidal correlation time to 10.04 seconds.

Graphs showing the increase in speed of the TMS320C25 over the
T800 for unpacked and packed pixels are given in figs 3 and 4
respectively. The T800 recorded times for array sizes 3x3 and 4x4
are slightly inaccurate due to the timing method used. If these
inaccuracies are ignored the average increase in speed of the
TMS320C25 over the T800 for packed pixels is 22.19 times faster using
integer arithmetic, and 21.32 times faster using real arithmetic.
For packed pixels, the increase in speed is reduced to 2.44 times
faster for integer arithmetic and 2.45 times faster for real
arithmetic.

These figures are for the T800 operating at 20MHz whilst the
TMS320C25 operates at 40MHz. It would appear that the TMS320C25 is
executing the correlation program twice as fast, giving it an unfair
advantage. However, the T800 has an instruction cycle time of 50ns
whilst the TMS320C25 has an instruction cycle time of 100ns;
therefore the T800 is actually executing program code twice as fast
the TMS320C25.

Using packed pixels on the T800 decreased the time required to perform the correlation. This is because the access to slower external memory has been reduced due to two pixels being read in one external read cycle. The correlation time is not halved because additional cycles are required for the unpacking operation.

A cycle-by-cycle analysis of both the TMS320C25 correlation programs was undertaken. For unpacked pixel data storage it was found that maximum program efficiency was obtained when using large array sizes. When using packed pixels the analysis found that the unpacking operation dominated the time required. This was because the unpacking operation requires an 8-bit shift right of the data; as the TMS320C25 can only shift right one bit at a time, a shift right bottleneck is produced. This is the cause of the reduced correlation performance.

The results have shown that the TMS320C25 is superior in performance to the T800 when performing correlation. As a large number of image processing functions use the MAC operation, the TMS320C25 is more suited to these operations than the T800 (as shown by the correlation results). However, as Transputers are designed to be linked together (rather than operate independently) an array of Transputers could be used to provide increased performance. It would require at least 23 Transputers to match the speed of the TMS320C25 when performing unpacked pixel correlation, and at least 3 Transputers when performing packed pixel correlation. These numbers would be increased in practice due to the communication bottleneck when using Transputer links.

Therefore, a digital signal processor has been chosen as the processing element in the image processing system which is being developed. The TMS320C25 will not be used because of its shift right bottleneck and its limited memory capacity (it cannot store two 512x512 images); instead the Texas Instruments TMS320C30[15] will be used. This device resolves the shortcomings of the TMS320C25 by containing a single cycle barrel shifter and a large address space. The TMS320C30 also contains a floating point multiplier/accumulator, an onboard DMA controller, more onboard memory, two serial ports and two timers.

9. IMAGE PROCESSING SYSTEM

The image processing system currently being designed will be comprised of a single processor board (called a Node) in its minimal form which can be expanded up to a total of eight Nodes. The Nodes will be connected together in a hypercube configuration. Each Node (see figure 5) will consist of two TMS320C30 processors, ROM, RAM and I/O peripherals. The processors will be operated at 33MHz which enables a MAC operation to be performed in 60ns. Each processor will have 8K by 32-bits of EPROM which will be used to boot the processor and will contain a toolbox of commonly used routines. Read/write memory will consist of 512K by 32-bits of DRAM which is sufficient to store several 512 x 512 images. Application software will also be loaded into this memory together with the toolbox routines which will be loaded into DRAM from the slower EPROM on power up.

Communication between the two processors on the Node will be via 2K by 32 bit dual port RAM. Inter Node communication will be performed by means of the three serial links on each Node. Each processor also contains an Inmos Link adapter which will enable communication with Transputer devices. This is necessary because the image acquisition and display system is based upon a Transputer board. Images will be transferred between the Transputer system and a Node via Transputer links. An RS232 port is provided to enable connection to a terminal. Parallel communication to external devices is provided by a SCSI interface.

System software will include routines to download/upload images from the Transputer system and common image processing operations such as edge detection and filtering. These routines will be stored in the toolbox contained in the EPROM. After the low level software has been written, high level applications including correlation will be implemented. Investigations into the optimum partitioning of both the low and high level applications software will be made to ensure that the multiple processors are used efficiently.

10. CONCLUSIONS

It has been shown that correlation using a hierarchy of image resolutions significantly reduces the time required to perform image correlation. Uniprocessor systems are unable to provide sufficient processing power to produce correlation results in a reasonable time. To overcome this, multiprocessor systems can be used but it is important to choose the right processor. Results have shown that digital signal processors can produce superior performance over conventional microprocessors. An image processing system using multiple digital signal processors is currently being designed.

11. ACKNOWLEDGEMENTS

The authors are pleased to acknowledge the support of the Coherent and Electro-Optics Research Group, the National Advisory Body, and the Science and Engineering Research Council, for the work reported in this paper.

12. REFERENCES

1. Koukash, M.B.Q.S., PhD Thesis, Liverpool Polytechnic, 1987.

2. Anuta, P.E., "Spatial Registration of Multispectral and Multitemporal Digital Imagery using Fast Fourier Transform Techniques", Digital Image Processing for Remote Sensing, Bernstein, R., pp.122-137, IEEE Press, 1970.

3.   Flynn, M.J., "Very High-Speed Computing Systems", Proceedings of the IEEE, Vol. 54, pp.1901-1909, 1966.

4.   Schmitt, L.A. and Wilson, S.S., "The AIS-5000 Parallel Processor", IEEE transactions on Pattern Analysis and Machine Intelligence, Vol. 10, No. 3, pp.320-330, 1988.

5.   Batcher, K.E., "Design of a Massively Parallel Processor", IEEE Transactions on Computers, Vol. C-29, No. 9, pp.836-840, 1980.

6.   Hillis, W.D., The Connection Machine, MIT Press, 1986.

7.   Seitz, C.L., "The Cosmic Cube", Communications of the ACM, Vol. 28, No. 1, pp.22-33, 1985.

8.   Hayes, J.P. et al., "A Microprocessor-based Hypercube Supercomputer", IEEE Micro, Vol. 6, No. 5, pp.6-17, 1986.

9.   Barrett, D.J., "Image Processing and Video Rate I/O on Transputer Arrays", IEE Colloquium on Transputers for Image Processing Applications, Digest No. 1989/22, pp.4/1-4/7, IEE, 1989.

10.   McIlroy, C.D., Linggard, R. and Monteith, W., "Hardware for Real-time Image Processing", IEE Proceedings, Vol. 131, Pt. E, No. 6, pp.223-229, 1984.

11.   Ruetz, P.A. and Brodersen, R.W., "Architectures and Design Techniques for Real-time Image-Processing IC's", IEEE Journal of Solid-State Circuits, Vol. SC-22, No. 2, pp.233-250, 1987.

12.   Inmos, IMS T800 Transputer Data Sheet, Inmos Ltd., 1987.

13.   Texas Instruments, TMS320C25 User's Guide, Texas Instruments, 1986.

14.   Inmos, OCCAM 2 Reference Manual, Prentice Hall, 1988.

15.   Texas Instruments, Third-Generation TMS320 User's Guide, Texas Instruments, 1988.

Unpacked Pixels

T800 Image/Template Storage

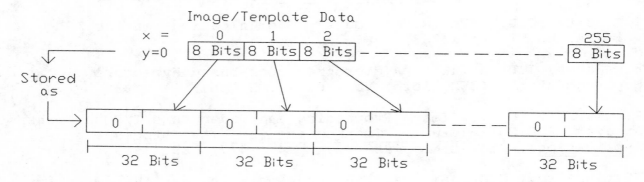

Figure 1a.

TMS320C25 Image/Template Storage

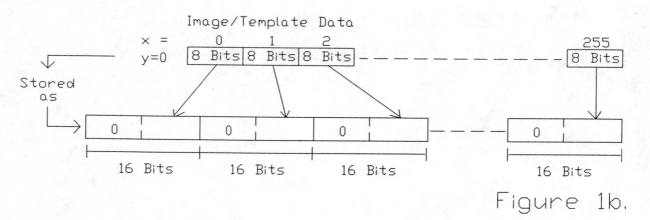

Figure 1b.

Packed Pixels

T800 Image/Template Storage

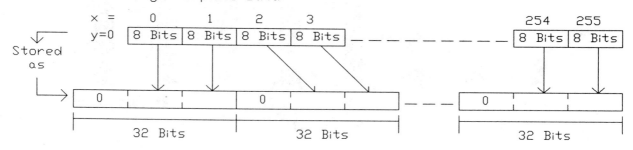

Figure 2a.

TMS320C25 Image/Template Storage

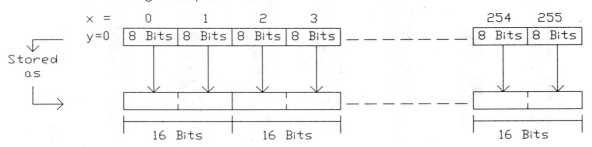

Figure 2b.

Unpacked Pixel Speed Increase of The TMS320C25 Over The T800

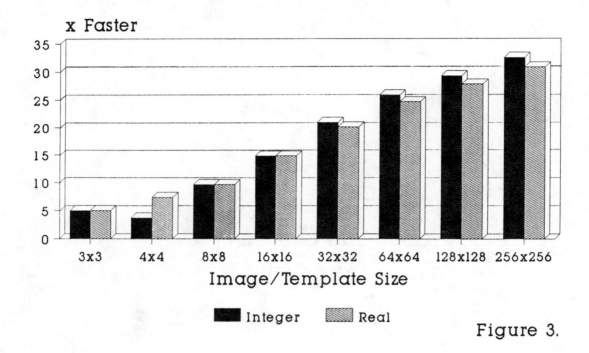

Figure 3.

Packed Pixel Speed Increase of The TMS320C25 Over The T800

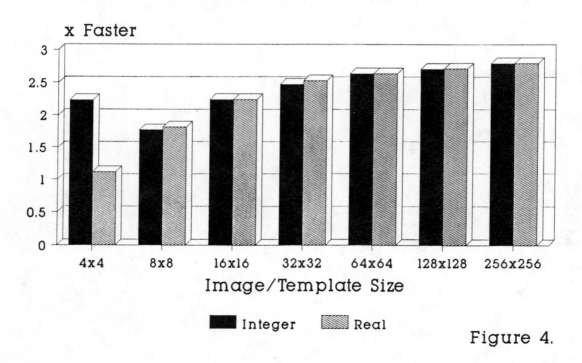

Figure 4.

TMS320C30 Based Dual Processor Node

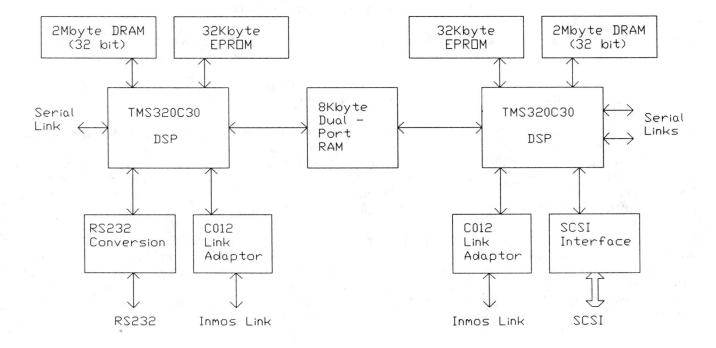

Figure 5

3-D DISPLACEMENT ANALYSIS USING OBLIQUE AXIS SPECKLE PHOTOGRAPHY

(1)   (2)   (2)
SMITH E W,   TAN Y S   &   HE Y M.

(1) Dept of Production Technology, Massey University,
Palmerston North, New Zealand

(2) Dept of Mechanical Engineering, Xi'an Jiaotong University,
Xi'an, Shaanxi Province, China.

ABSTRACT

It is usual, in speckle photography, to set the optical axis of the photographic lens so that it is perpendicular to the surface being investigated. However if the photographic lens is set with its optical axis at an angle to the object surface then it is shown that the accuracy of separating out the in-plane and out-of-plane displacements can be improved. A further benefit is that the whole visual field of the photographic lens can be studied whether the object itself is flat or has a 3-D surface profile. This means that speckle photography can be applied to objects of any shape and of considerable size.

Firstly the principle and theory of the oblique axis method is explained. In particular the method of correct focusing is shown; because of this the technique is not possible with simple 35 mm SLR cameras where lens and film planes do not have the necessary movements for differential focusing, none the less the potential of a mono-rail plate camera can be full realized. Secondly the validity of the technique will be seen by presenting results of tests on a simple structure. Finally it will be shown, using automatic fringe analysis, that the method can be applied to large 3-D structures, such as a machine tool.

1.   INTRODUCTION

Archbold and Ennos [1], in their comprehensive practical study of laser speckle photography, considered the case of viewing with the optical axis of the photographic lens normal to the object surface. They looked at all possible object displacements: lateral translation, in-plane rotation (being a particular case of lateral translation), out-of-plane translation and tilt (see Fig 1).

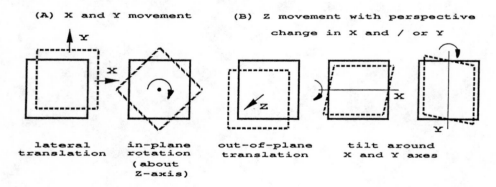

Fig. 1   Separation of 3-D displacements using speckle photography.

It was shown how out-of-plane translation and tilt could be measured; they also showed that both could contribute errors to the measurement of lateral translations. Ennos and Virdee [2] later used defocused laser speckle photography to examine the curved surface of a cylinder subjected to internal pressure; by careful consideration of certain conditions of defocus they applied the technique to measure out-of-plane displacements. Compared with holographic interferometry, processing data from two speckle photographs was rather cumbersome. From the early work of Ennos and co-workers it has always been accepted that the laser speckle method is best suited for measuring in-plane or lateral displacements over flat surfaces. This is borne out by the practical survey of applications of the laser speckle method carried out by Chiang, et al. [3]. However with the white light speckle method there is no complementary white light technique for measurement of out-of-plane displacements; thus there is much interest in using the technique to look at 3-D displacements [4].

Asundi and Chiang [5] highlighted three techniques to separate out 3-D displacements using the white light method: firstly, if in-plane movement is small compared with that out-of-plane (in the case of bending, for instance), then a lens of relatively small focal length and minimal magnification will suppress the effect of in-plane displacements and the result will be out-of-plane sensitivity only; secondly, exposing two specklegrams from two different X coordinates can lead to the separation of d_x and d_z, a third exposure using a different Y coordinate will lead to the separation of d_y and d_z; and thirdly, combining the specklegram with moiré grid projection to form a speckle-moiré-gram, which with full-field filtering allows d_z to be found from moiré information and d_x and d_y from the speckle information. None of these techniques is ideal: the first is restricted in the main to bending; the second requires exposing three speckle photographs; the third is restricted to flat surfaces with normal illumination. Vikram and Vedam [6] showed that the three speckle photographs of the second method must be increased to four if the signs of displacement are not known. But if a reference translation is introduced between exposures then only two negatives are needed [7]. As well as resolving the question of sign and reducing the number of specklegrams to be exposed, the reference translation also increases the sensitivity of the method; normally the minimum sensitivity is the speckle diameter and in the case of white light speckle the size of the characteristic speckle is often much greater than with laser speckle.

From the above brief survey if can be seen that the technique of speckle photography (laser and white light) has proved to be an excellent method for establishing 3-D displacements. However one limitation which does restrict its use is the requirement to make viewing normal to the surface. If the surface under strain is at an angle to the optical axis of the photographing lens then differential focusing occurs across the field of view; furthermore the image will undergo differential magnification. For small objects it is usually quite practical to set up the viewing camera so that it is normal to the subject under consideration; but with large objects where optimum magnification is sought it may not be practical. The consequence is oblique viewing. This dictates a need for differential focusing by orientating the optical axis to the image plane.

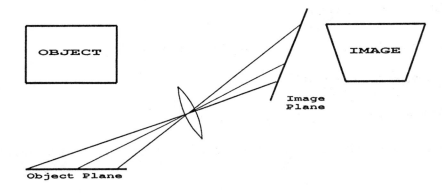

Fig. 2   Differential magnification due to oblique viewing.

This is standard practice when using a plate camera in pictorial photography with full movements but the usual convenience of the 35 mm camera (as used by Archbold and Ennos in their original study) does not go so far. Thus there is a need for tilt of the image plane about the X and/or Y axes. Differential magnification of the image resulting from oblique viewing is depicted in Fig. 2

In this paper the theory of oblique axis viewing will be developed to show that the phenomenon of differential magnification can be accounted for in calculating displacements, and separation of the in-plane and out-of-plane displacements can be made. As with the method of Benckert, Jonsson and Molin [7] it is necessary to make specklegrams at two separate positions and to give a reference translation between exposures to allay any ambiguity of displacement direction.

2. THEORY OF THE OBLIQUE SPECKLE PHOTOGRAPHY METHOD

2.1 Focusing principle

When the Oblique Speckle Photography method is used the optical axis is not perpendicular to the object plane being measured so that it is necessary to establish correct focusing of the surface plane. Fig. 3 shows the relationship between object surface plane, focusing lens and image plane. θ is the angle between the normal to the surface plane and the optical axis (which is rotated in the horizontal plane only), the point A is the intersection of the surface with the optical axis, A' is the corresponding point on the image plane and B is an arbitrary point on the object surface. For correct focusing of the object surface the image plane is rotated to angle θ_1; this angle is fixed under correct focusing.

From the definition of magnification let,

$$M_1 = \frac{A'O}{AO} \tag{1a}$$

$$M_2 = \frac{B'O}{BO} = \frac{DO}{CO} \tag{1b}$$

and

$$CO = AO - AB. \sin \theta$$

$$DO = A'O + DA' = M_1.AO + M_2.AB. \cos \theta . \tan \theta_1$$

Substituting the above expression into (1b) gives

$$AO.(M_1 - M_2) = -M_2.AB. (\cos \theta . \tan \theta_1 + \sin \theta) \tag{2}$$

Using the Gaussian lens formula, the image distances can be related to the focal length, f

$$A'O = \frac{f.AO}{(AO-f)} \tag{3a}$$

$$DO = \frac{f.CO}{(CO - f)} \tag{3b}$$

Substituting (1) and (3) into (2) and simplifying gives

$$\tan \theta_1 = \frac{f.\tan \theta}{(AO - f)} \tag{4}$$

This expression shows that for a set object distance, focusing on the image plane depends on the angle between the optical axis and the object plane for a given lens. Further, when $\theta = 0$ the expression reduces to the situation of the optical axis being perpendicular to the object surface.

2.2 Calculation of the magnification

In Fig. 3 the co-ordinate plane XOY is parallel to the object surface. If the optical axis has rotation θ in the XOY plane, then the magnification of an arbitrary point on the surface will be a function of x. The magnification of the arbitrary point B will be,

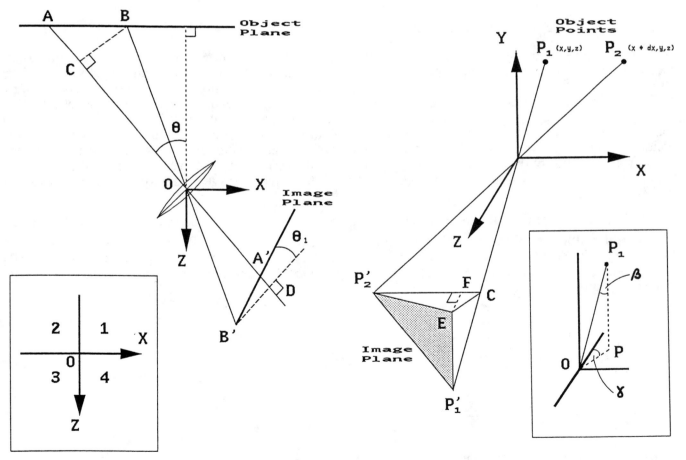

Fig. 3   Oblique focusing.   Fig. 4   Object points and the image plane.

$$M_b = \frac{f}{CO - f}$$

where $\quad CO = AO - (x_b - x_a) \cdot \sin \theta$

giving $\quad M_b = \dfrac{f}{Ao - f - (x_b - x_a) \cdot \sin \theta}$ (5a)

when B has the general co-ordinate position (x,y,z), then (5a) becomes,

$$M = \frac{f}{|z| \cos \theta - f - (x - x_a) \cdot \sin \theta}$$ (5b)

The sign of θ is determined by use of the inset in Fig. 3 where, if the position of the optical axis is in the 1 and 3 quadrants of the ZOX co-ordinate plane the sign is positive, otherwise it is negative.

2.3   Correction of the in-plane displacement formula

When calculating the in-plane displacements on the object surface it is necessary to make a correction to the speckle displacements on the image plane due to the angle between image plane and object surface. The oblique angle of the optical axis in the horizontal plane makes magnification a function of x: thus a straight horizontal line on the object surface, ie x direction, will become an oblique line on the image plane. This is highlighted in Fig 2. Thus when a point on the object has a displacement in the x direction, it will contribute to the displacement in the y direction on the image plane. The contribution will depend on the oblique angle of viewing; but only indirectly on the focal length of the lens, in as much as it affects the magnification. Shift of the image plane to conpensate for the converging lines can be exploited but this will then lead to out of focus areas when magnification is high. Loss of image area may also occur if extreme shift is required.

In Fig. 4 point O is the optical centre of the imaging lens. $\Delta P_2' E \, P_1'$ is on the image plane. P_1 and P_2 are points on the object surface and C is the intersection of the straight line $P_1 P_1'$ with the plane which is parallel to the XOY co-ordinate plane and goes through the point P_2'. Line $P_2'C$ is parallel to the line P_1P_2 which connects two points on the object surface. If M_1 and M_2 represent the magnifications at points P_1 and P_2 respectively, then

$$M_1 = \frac{OP_1'}{OP_1} = \frac{CP_1' + OC}{OP_1}$$

$$M_2 = \frac{OP_2'}{OP_1} \quad \text{or} \quad \frac{OC}{OP_1}$$

giving $\quad CP' = M_1 \cdot OP_1 - OC = (M_1 - M_2)\, OP_1$

and
$$CE = P_2'E.\tan(\theta + \theta_1)\cos\gamma$$

$$EP' = \frac{CE}{\tan\beta} = \frac{y.P_2'E\tan(\theta + \theta_1)}{|z|} \tag{6}$$

$$P_2'C = P_2'E.[\cos(\theta + \theta_1) + \sin(\theta + \theta_1).\tan\gamma]$$

$$= P_2'E[\cos(\theta + \theta_1) - \frac{x.\sin(\theta + \theta_1)}{|z|}] \tag{7}$$

Equations (6) and (7) can be rewritten for clarity by including factors which depend only on the focusing system,

$$EP_1' = K_y. P_2'E \tag{8a}$$

$$P_2'C = K_x. P_2'E \tag{8b}$$

where the geometric factors

$$K_y = \frac{y.\tan(\theta + \theta_1)}{|z|} \tag{9a}$$

and
$$K_x = \cos(\theta + \theta_1) - \frac{x.\sin(\theta + \theta_1)}{|z|} \tag{9b}$$

EP_1' is the contribution of $P_2'E$ to the y direction and as such will give rise to an error in the measured displacement in the y direction. EP_1' needs to be subtracted from the recorded displacement in the y direction.

3 POINTWISE FILTERING AND OBLIQUE SPECKLE PHOTOGRAPHY

If the speckle photographs are analysed by the point-by-point filtering method, then the in-plane displacements are calculated as follows:

$$d_x = \frac{K_x\cos\alpha\,\lambda\,L}{MS} \tag{10a}$$

and
$$d_y = \frac{(\sin\alpha - K_y\cos\alpha)\lambda L}{MS} \tag{10b}$$

where, as shown in Fig. 5, L is the distance from the speckle photograph to the Young's fringes imaging screen, λ is the laser wavelength used for the filtering, S is the spacing of the fringes and α is the orientation of the fringes. It will be noted that M will now vary over the image area.

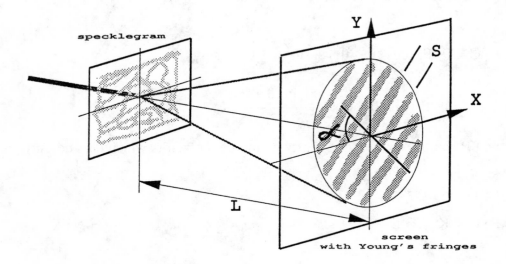

Fig. 5   Set-up for point-by-point filtering.

When normal viewing is used $\theta = 0$ and K_x will become unity and K_y zero; then the above equations reduce to the standard equations.

To measure 3-D displacements using Oblique Speckle Photography it is necessary to make two speckle photographs which are taken from different positions and exposed simultaneously. Then if d_x, d_y and d_z are the displacements along the orthogonal axes with reference to the object it is necessary, for their evaluation, to write the equation in matrix form, viz.

$$(11)$$

$$
\begin{bmatrix}
1 & 0 & \dfrac{x_I}{|z_I|} & 0 \\[2ex]
0 & 1 & \dfrac{y_I}{|z_I|} & 0 \\[2ex]
1 & 0 & \dfrac{x_{II}}{|z_{II}|} & 0 \\[2ex]
0 & 1 & \dfrac{y_{II}}{|z_{II}|} & 0
\end{bmatrix}
\begin{bmatrix}
d_x \\[2ex] d_y \\[2ex] d_z \\[2ex] 0
\end{bmatrix}
=
\begin{bmatrix}
d_{x_I} \\[2ex] d_{y_I} \\[2ex] d_{x_{II}} \\[2ex] d_{y_{II}}
\end{bmatrix}
$$

Where subscripts I and II refer to separate camera positions. The square matrix refers to the co-ordinate positions of points on the object surface with respect to origins at the camera lens. The column matrix on the right lists the displacements in x and y directions taken from the point by point filtering of the speckle photographs for each camera position, ie

$$d_{x_i} = \frac{K_{x_i} \cos \alpha_i \, \lambda \, L}{M_i S_i}$$

(12a)

and

$$d_{y_i} = \frac{(\sin \alpha_i - K_{y_i} \cos \alpha_i) \, \lambda \, L}{M_i S_i}$$

(12b)

where $\qquad$ $i = I, II$

When the object displacement is less than the characteristic speckle size it is usual to subject the object to a known rigid body movement which is greater than the speckle size. This is equally feasible with the oblique method and the rigid body displacements may be subtracted to give the true values of d_{xi} and d_{yi}.

4. APPLYING THE METHOD TO A SIMPLE STRUCTURE

The method of Oblique Speckle Photography was applied to a simple cantilever subject to tip loading angled in the X-Z plane. The cantilever beam of length 215 mm is shown in Fig. 6 and two cameras are set up to measure displacements in the X and Z directions. White light illumination is provided from two slide projectors and the speckle pattern is generated by coating the surface with retro-reflective paint.

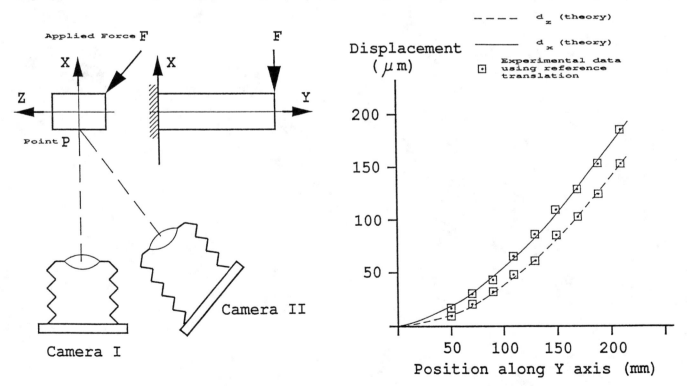

Fig. 6 Cantilever beam and oblique viewing. Fig. 7 Verification of results for bending of cantilever.

The cameras were placed to give nominal magnifications of point P, at the end of the cantilever on the X face, of 0.27 with Camera I and 0.25 with Camera II. Firstly, no reference translation was applied; the sign of deformation was then assumed. The smallest displacement that could be measured was restricted to 50 μm, the characteristic size of the glass particles in the reflective coating. Then in a second experiment a reference shift was applied to each camera support in the direction of the applied force on the cantilever.

This allowed displacements of the order of 10 μm to be measured as well as verifying the direction of displacement. Fig.7 shows the correlation between theoretical data and the experimental results.

5. APPLYING THE METHOD TO A COMPLEX STRUCTURE

Like holographic interferometry, speckle photography comes into its own as a displacement or strain measuring technique when full-field data is required and in particular when full-field displacements or strains are needed over large objects. Both techniques will highlight areas of high strain gradient which otherwise could be missed if conventional point-by-point techniques are used such as electrical resistance strain gauges. Thus in considering the workability of the Oblique Speckle Photography method it is important to consider its application to large structures. White light Oblique Speckle Photography was used to measure the 3-D thermal deformation of the head-stock of a machine tool lathe. The experiment was carried out in a standard machine shop. The lathe head-stock was painted with retro-reflective silver paint (specially formulated to be peeled from machine surfaces after experimentation) and illuminated by two slide projectors. A very simple experiment was performed which works well when testing the feasibility of holography for strain measurement in machine tool structures - the lathe was run for two hours with the motor under no load, speckle photographs were exposed prior to running the lathe and then double exposed at the end of the run. Heating of the lathe structure will occur due to running of the motor, shafts, bearings and gears and the resultant thermal distortions of the heavy structure will be of the order of 100 micrometres or so. Results from point-wise filtering of two specklegrams are shown in Fig. 8. These were evaluated using an image processing system as shown in Fig. 9.

Ennos and Virdee [2] showed that laser speckle photography is capable of measuring out-of-plane displacement with an absolute accuracy approaching that of holographic interferometry. Also, holographic interferometry is insensitive to movement normal to the line of sight, so that to measure 3-D displacements requires the use of three strategically placed holograms.

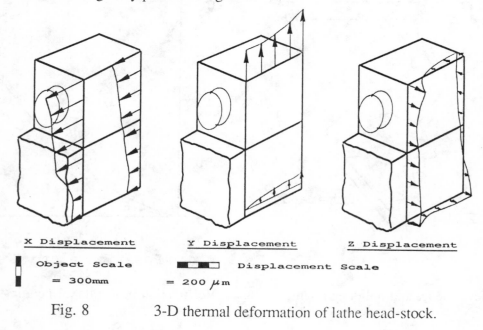

X Displacement   Y Displacement   Z Displacement

■ Object Scale   ■■□□ Displacement Scale
= 300mm   = 200 μm

Fig. 8   3-D thermal deformation of lathe head-stock.

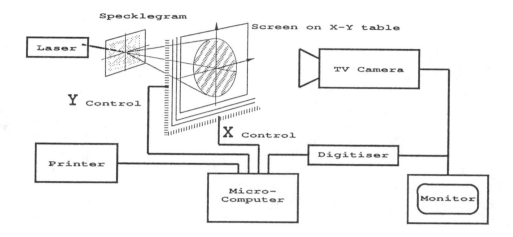

Fig. 9   Automatic point-by-point filtering by image processing.

Tan and Smith [8] have shown that holography can be used to measure static and dynamic displacements in machine tool structures and typical displacement fringes resulting from the thermal distortions of a milling machine tool are shown in Fig. 10. It will be necessary to process the fringe patterns from a set of three hologram plates to establish 3-D displacements; this will necessitate the same amount of image processing and computation as with the speckle photography method. However the complexity of the set-up for holography is far greater than that for speckle photography; a laser with two metres or more coherence length is not required for speckle photography, even if laser speckle is employed. Furthermore vibrational stability whilst exposing the speckle photographs is not especially critical as is the case with holographic interferometry.

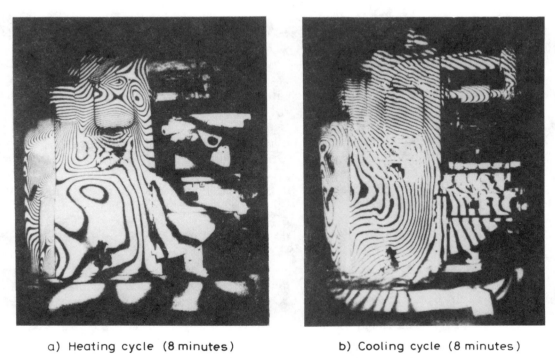

a) Heating cycle (8 minutes)   b) Cooling cycle (8 minutes)

Fig. 10   Holographic fringes due to thermal distortions of a milling machine.

6. CONCLUSIONS

The Oblique Speckle Photography method has been introduced for analysing 3-D displacements in structures where it is not possible or desired to view the object normal to the optical axis of the camera lens. The method can be used for both laser and white light speckle photography. It can be applied to small or large structures and uses camera movements typical of the standard mono-rail plate camera. Two speckle photographs must be taken from separate positions and a reference translation introduced between exposures; this allows 3-D analysis with full assessment of the signs of displacement, it also increases the sensitivity of the method. Image processing methods allow quick assessment of the Young's fringe patterns and calculation of displacements or strains. It is felt that the technique is comparable with holographic interferometry when looking at deformations in large structures, but does exhibit one important practical advantage: no special vibration isolation is necessary during photographic exposure.

7. ACKNOWLEDGEMENTS

The first mentioned author would like to thank Professor Ku Chun Hsian of the Department of Mechanical Engineering, Xi'an Jiaotong University for a very rewarding study leave spent with his staff and department. Some aspects of this paper have arisen from that visit. All the authors would like to thank the staff of Xi'an Jiaotong University - too numerous to mention all by name - who have helped in this work.

8. REFERENCES

1   E Archbold and A E Ennos, "Displacement Measurement from Double-Exposure Laser Photographs", Optica Acta, vol. 19, pp 253-271, 1972

2   A E Ennos and M S Virdee, "Laser Speckle Photography as a Practical Alternative to Holographic Interferometry for Measuring Plate Deformation", Optical Engineering, vol. 21, pp 478-482, May/June 1982.

3   F P Chiang, J Adachi, R Anastasi and J Beatty, "Subjective Laser Speckle and Its Application to Solid Mechanics Problems", Optical Engineering, vol. 21, pp 379-390, May/June 1982.

4   A Asundi and F P Chiang, "Theory and Applications of the White Light Speckle Method for Strain Analysis", Optical Engineering, Vol. 21, pp 570-580, July/August 1982.

5   A Asundi and F P Chiang, "Separation of 3-D Displacement Components in the White Light Speckle Method", Optics and Laser Technology, Vol ?, pp 41-45, February 1983.

6   C S Vikram and K Vedam, "Complete 3-D Deformation Analysis in the White Light Speckle Method", Applied Optics, vol. 22, pp 213-214, 1983.

7   L Benkert, M Jonsson and N E Molin, "Measuring True In-Plane Displacements of a Surface by Stereoscopic White-Light Speckle Photography", Optical Engineering, vol. 26, pp 167-169, February 1987.

8   Y S Tan and E W Smith, "Measurement of the Static, Dynamic and Thermal Rigidity of Machine Tool Structures Using Holography", Proc. 22nd Machine Tool Design and Research Conf., Manchester, UK, pp 159-171, 1981.

ACCURACY OF FRINGE PATTERN ANALYSIS

GUI-YING WANG   XIAO-PING LING

SHANGHAI INSTITUTE OF OPTICS AND FINE MECHANICS,
ACADEMIA SINICA, P.O.BOX 8211

ABSTRACT

In this paper, affecting factors on accuracy of fitting wavefron of light beam with orthogonal polynomials are given theoretically in terms of F-test method of statistics. The method availability to control accuracy has also been verified experimentally by an axial hologram reconstruction.

I. Introduction

Since Saunders first proposed quantitative analysis for shear interferogram[1], the processing methods have been developed greatly with mordern computer. The automatic processing of interferent fringe is convenient and precise. In interferent fringe analysis, according to principle of forming interferent fringe, a wavefront function $F(X_k, Y_k)$ is consisted of the degree and the sample points of the interferent ringes. There are two methods of selecting sample point. In the first method, the sample points can be equal distance and the wavefront function is consisted of data-fitting with the real sample poins through a matrix that was known[2,3]. The accuracy depends on precision of determining the degree numbers of the fringes. Hence it is inconvenient and wasting of CPU time. In the second method, sample points can be nonuniform distance and fitting wavefront may be taken orthogonal polynomials interpolation[4,5]. Here is only need to count locations of maximum intensity of the fringes, and the computation of determining the degree of the fringes can be omitted. So that accuracy of fitting wavefront is improved and CPU time is shortened. Although the method has been used widly and improved, there are some argumentative conception, for example, how to solve relationship of the fitting accuracy with the number of sample points and the degree of the orthogonal polynomials. In general, the better accuracy the more sample points. But, how many degrees of the polynomials are appropriate as in the case of certain accuracy and certain sample points, and how to control accuracy still key point of saving time of CPU time and so on. These will be discussed in the paper.

Takeda et al. also proposed a Fourier transform method of fringe pattern analysis[6] and it was further simplified and improved by using the sine function proposed by Mertz [7]. The advantage of the method is not need to find location of corresponding peak intensity of the fringes, so that affect of nonuniformly radiation of accuracy fitting is reduced. The method is suitable, particularly for automatic processing of moire topography. The heterodyne and homodyne of interferency and the fringe scanning technique have been developed [8]. These mothed can be processing in real time, but these are not subjects in this paper.

2. ACCURACY CONTROL OF FITTING WAVEFRONT

Reference [5] gave an example of fitting wavefront using orthogonal polynomial interolation at nonuniform sample points. If the degree number V_k of interferent frige is taken as an independent parameter, the orthogonal polynomial elements is taken as a variable and suppose that δ_k is the difference between an experiment value and a regression value. This is a linear regression problem because the sample points are limited. So that the stochastic error is governed by the normally distributed and the optimum linear error is evaluted by the least–quares method.

$$\delta_k^2 = \sum_{k=1}^{m} \left\{ \left| (V_k - F(X_k, Y_k)) \right|^2 \cdot W(X_k, Y_k) \right. \tag{1}$$

$$F(X_k, Y_k) = \sum_{i=1}^{N} C_\ell \, \mathcal{G}_\ell (X_k, Y_k) \tag{2}$$

where $W(X, Y)$ is arbittary positive weight factor, as parameters choised are appropriate, $W(x, y) = 1$. The (x, y) express the orthogonal polynomial to combinate $F(x, y)$. If $F(X, Y)$ is directly taken as order series of x and y, the matrix could be not computated because it involve Hilbert's matrix [9]. Hence one used Forsythy's method to make (X, Y) that is an that is an orthogonal polynomials consisted of the order series of x and y [10]. The coefficients {C} of the polynomials can be solved from (1) and also depende on the orthogonal property of them. Because wavefront function of the two dimension can be trasformed as one dimension to process [5], so we suppose

$$\mathcal{G}_\ell (X_k) = \sum_{i=1}^{\ell} a_{i\ell} X_k^{i-1} \cdot W(X_k) \tag{3}$$

then

$$a_{i\ell} = \sum_{k=1}^{m} X_k^{i-1} \cdot \mathcal{G}_\ell (X_k) \cdot W(X_k)$$

$$a_{\ell\ell} = \sum_{k=1}^{m} X_k^{\ell-1} \cdot W(X_k) - \sum_{i=1}^{\ell-1} a_{i\ell}^2$$

$$\mathcal{G}_\ell (X_k) = \frac{1}{a_{\ell\ell}} \left(X_k^{\ell-1} - \sum_{i=1}^{\ell-1} a_{i\ell}^2 \, \mathcal{G}_i (X_k) \right) \tag{4}$$

$$C_\ell = \sum_{k=1}^{m} \mathcal{G}_\ell (X_k) \cdot V_k \cdot W(X_k) \tag{5}$$

Let us using the F-test method to determine the optimum degree of the polynomials. Rewrite (1) as follows

$$V_k = F(x_k, y_k) + e_k \tag{6}$$

here e_k is a component of the error vector at the location (X_k, Y_k) and dependens only on the coefficients of the polynomials. We take it as a random variable, but $\mathcal{G}_\ell (X, Y)$ in the (2) is taken as an independent variable. The mean square of the sample book as follows, if

$$U_{il} = \sum_{k=1}^{m} \varphi_i(X_k) \cdot \varphi_l(X_k) \tag{7}$$

$$Q_l = \sum_{k=1}^{m} V_k \cdot \varphi_i(X_k) \tag{8}$$

let (7) and (8) instead of (1)

$$\delta_l^2 = \sum_{k=1}^{m} V_k^2 - \sum_{i=1}^{l} U_{ii} Q_i \tag{9}$$

$$\delta_{l+1}^2 = \delta_l^2 - C_{l+1}^2 \tag{10}$$

let

$$S_l^2 = \delta_l^2 / (m-l) \tag{11}$$

$$F = S_i^2 / S_j^2 ; \qquad j = i + 1 \tag{12}$$

where m shows maximum order number of the sample points and l shows the degree numbers of fitting polynomials. And $U_{ii} = 1$ and $U_{il} = 0$. We have computed S_l and S_{l+1} separately, then according to the statistical criterion judge them if coming from the same generaive book. If choosing the probabity value is 5% and the sample points are 64*64, then F=1.53 [11]. If while probabity value as the same value mentioned above and the sample points are 128*128, then F=1.34. As $S_l / S_{l+1} > F$, it is shown that S_l and S_{l+1} come from different generative book, so that we should choose $l + 1$ degree of the polyomial and continue to compare with next degree of the polynomial until $S_{l+1+h} / S_{l+2+h} < F$. But if $S_l / S_{l+1} < F$, it will be shown that S_l and S_{l+1} generate in the same same generative book. Hence the l degree of orthogonal polynomials can be used.

Table (1) shows S_l and F_{com} of an axial hologram. The computating results are the same order as taking the degrees about 5 and taking about 22 of the polynomials as sample points are 128*128. We consider if it is the best to take 15 degrees of the polynomials because both satisfactive accuracy and saving CPU time are important. The accuracy of the fringe analysis may be up to /50.

3. EXPERIMENTAL EXAMINATION

Figure 1 shows a scheme of the axial holographic interferometer which involves a telescope of enlargement factor M [12]. If wavefront curvature radius of reconstructed laser beam is R', then the real radius is

$$R_{real} = M^2 (1 + \frac{l}{M^4}) R' \frac{\lambda_{reco}}{\lambda_{real}} \tag{13}$$

where λ_{real} is the wavelength of measured laser beam and λ_{reco} is the wavelength of reconstructing laser beam. Figure 2 shows the axail hologram which is

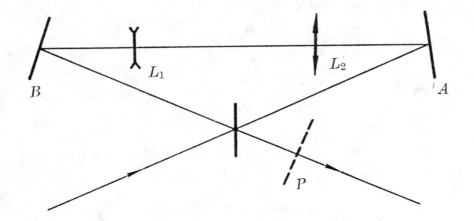

Fig.1. Scheme of the axial holographic interferometer

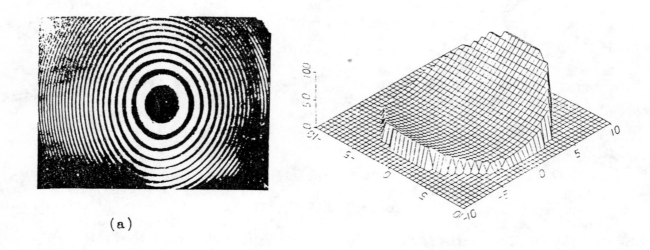

(a)

Fig.2. A axial hologram of the high power laser beam and
its wavefront depending the fringes analysis.

also a radial shear inteferogram. Using holographic reconstruction the wavefront radii are given, R1=5.64m and R2=8.00m corresponding to astigment aberration separately. The average wavefront radius

$$\frac{1}{R} = \frac{1}{2}\left(\frac{1}{R_1} + \frac{1}{R_2}\right)$$

(14)

It is obtained, R =54.51m, where the accuracy can be up to 1%[13]. On the other hand, using interferent fringe automatic processing can also yield the wavefront radius. It is well know that the orthogonal polynamials of fitting wavefront can be transformed into Zernike's polynomial [13].

$$W(\rho, \theta) = C1 + C2 \cdot \rho \cdot \sin\theta + C3 \cdot \rho \cdot \cos\theta + C4 \cdot \rho \cdot \sin\theta + C5(2\rho - 1) + C6 \cdot \rho \cdot \cos\theta$$

(15)

The total average wavefront radius can be yield by optimal reference sphere concept [14]. The phase difference of the reference sphere with a radius of R is expressed

$$W(\rho, \theta) = A(\rho, \theta) + D\rho^2 + K \cdot \rho \cdot \sin\theta + L\rho\cos\theta + M$$

(16)

Comparing (15) and (16), that are yirlded

$$D = -2C5$$

$$K = -C2$$

$$L = -C3$$

$$M = -C1$$

(17)

Table 2 shows coefficients of the first 5 degrees of Zernike's polynomial. The normalizing radius of the reference sphere is

$$R_{Nor} = -\frac{1}{4C_5}$$

(18)

Here R =53.71m. The relatival error of both difference methods is E < 1.5% .

3. CONCLUSION

The Shear interferometry is an usual method in the interferometry. Using orthogonal polynomials interpolation can complete quickly and precisely to fit wavefront of a light beam. The accuracy depends on the numbers of sample points and the degrees of taking polynomials. As sample points have been determined, fitting acuracy can be controlled by F-test method of statistics. We first choosed probability value, and then choose the degree of the ortho-gonal polynomials depending on both the wavefront form to evaluate initially and Zernike's polynomial because the physical meaning of the polynomial is very clear and definite. The method of the control accuracy is not only secure well precision, but also saving time of computation.

Authors thank S. D. Wu for his helpful discussions in the interferent fringe processing. We also would like thank Qiau Jia Gin and Zheng Yan Ling for their helping to operate computer and to draw picture.

Table.1 Coefficients of Zernike's polynomial corresponding Fig.2.

C1	C2	C3	C4	C5	C6
0.000698	0.000011	0.000090	0.000032	0.000639	0.000174

Table.2 The error square S1 of fitting the wavefront

N	S1	Fcom
1	3.7524E−9	
2	2.5242E−9	1.4866
3	1.5680E−9	1.6098
4	3.3112E−11	473.54
5	3.3600E−11	0.9854
6	3.0806E−11	1.0907
7	3.1249E−11	0.9858
8	2.5748E−11	1.2136
9	2.6150E−11	0.9846
10	2.6409E−11	0.9902
11	2.6548E−11	0.9948
12	2.3800E−11	1.1156
13	1.2021E−11	1.1156
14	1.2224E−11	1.9790
15	1.2215E−11	0.9833
16	1.2405E−11	0.9847
17	1.1351E−11	0.9855
18	1.1533E−11	1.0929
19	1.1596E−11	0.9842
20	1.1815E−11	1.9815
21	1.1951E−11	0.9886
22	1.0325E−11	1.1444
23	9.5853E−12	1.0772
24	9.7783E−12	0.9803
25	9.9814E−12	0.9797
26	1.0191E−11	0.0981
27	1.0412E−11	0.9788
28	9.2911E−12	1.0567
29	8.7186E−12	1.0373
30	8.4047E−12	1.0230
31	8.2154E−12	0.0985
32	8.3958E−12	0.9987
33	8.5862E−12	0.9782
34	8.7774E−12	0.9735
35	9.0084E−12	1.0295
36	8.6664E−12	1.0395

4. REFERENCE

1. I. B. Saunders; 'Measurement of wave fronts without a reference standard', J.Res.Bur. tand, Vol.65B, No.4,239,1961.

2. M.P.Rimmer; 'Mothod For Evaluating Lateral Shearing Interferograms', Appl. Opt, Vol.13, No.3, 623, 1974.

3. D. Malacara and M. Mendez; 'Lateral Shearing Interferometry of Wavefronts having rotational symmetry', Opt. Acta, Vol.15. No.1, 59, 1968.

4. D. Dotton, A. Cornezo and M. Latta; 'A Semiautomatic Method for Interpreting Shearing Interfograms', Appl. Opt., Vol.7, No.1, 125, 1968.

5. Wu Shudong and heng Hui; 'Interferogram Analysis With Microcomputer', Acta Optica sinica Of China, Vol.3, No.9, 815, 1983.

6. M.Jakeda et al.; 'Fourier-transform Method Of Fringe-pattern Analysis For Computer-based Topography And Interferometry'. Vol.72, No.1, 156, 1982.

7. L. Mertz; 'Fourier-transform Method of Fringe-pattern Analysis For Real-time Fringe Pattern Analysis'. Vol.22, No.10, 1535, 1983.

8. J. Bruning et al. 'Digital Wavefront Measuring Interferometer For testing optical Surface And Lenses', Appl. Opt. Vol.13, No.11, 2693, 1974.

9. D.Hilbert,Ein Beitray Zur; 'Theoric des Legendre'schen Polynoms', Acta.Math.Vol.18, 1804, 155.

10. G.E.Forsythe; 'Generation and Use of Orthogonal Polynomials for Data-Fitting With A Digital Computer', J.Soc.Indust. Appl. Math, Vol.5, No.2, 74, 1957.

11. Zhou Gai-Rong; 'Probability and Mathematical Statistics',Publishing House of Higher Education, 1984.

12.en-yang Yu and Gui-ying Wang; 'Holographic Diagnosis Of Wavefront of a laser beam', Acta Optical sinica of China, Vol.2, No.4,349,1982.

13. D.A.Malacare; 'Optical Shop Testing', John Wiley & Sons. New York, 1978.

14.K.P.Kiewicz; 'Determinetion Of The Optimal Reference Sphere', J. Opt. SOC.Am, Vol.69, No.7, 1045, 1979.

ADDENDUM

The following paper was presented at this conference, but the manuscript supporting the oral presentation is not available.

[1163-17]   **Automatic noncontact surface profiler**
M. Koukash, C. A. Hobson, Liverpool Polytechnic (UK)

AUTHOR INDEX

FRINGE PATTERN ANALYSIS

Volume 1163